CELESTIAL NAVIGATION
FOR YACHTSMEN

Celestial Navigation
for
Yachtsmen

FIFTH EDITION

Mary Blewitt, F.R.I.N.

EDWARD STANFORD LONDON

Edward Stanford Limited
Member of the George Philip Group
12–14 Long Acre London WC2E 9LP

First published 1950 for Yachting World
Fifth Edition 1971
New Impression 1972
Reprinted 1973
© Mary Blewitt 1971

Printed in Great Britain by
Hazell Watson & Viney Limited Aylesbury Bucks

ISBN 0 540 00354 9

FOREWORD TO FIRST EDITION
By Captain (E) J. H. Illingworth, R.N. (Ret.)

I am glad to be able to contribute a short foreword to this book because, having studied the text, I am sure that it will satisfy the very real need for an up-to-date work based on the use of the simplified tables and written entirely from the yachtsman's, as opposed to the airman's or ship navigator's, point of view.

Rigorous exclusion of unnecessary material from the text has, in my view, added to the value of the book and made it easier to attain that most desirable thing—correct emphasis on the various aspects of the subject. The result is lucid and practical, but there is just sufficient theoretical explanation included to inculcate the knowledge that is necessary as a basis for this subject.

PREFACE

Since the first edition of this book in 1950, a very large change has taken place in the field of navigation; while electronic aids have enormously lessened the reliance that big ships and aircraft put on celestial navigation, the increase in popularity of yachting has meant that a much larger number of amateurs need 'astro' to find their way about the seas. There used to be a tendency among navigators to treat celestial navigation as black magic and to imply that it was really too difficult for an ordinary chap to understand, this attitude has now disappeared and the increased teaching of mathematics in schools has produced a generation much less afraid of figures than the previous one.

This book is written for beginners and I have presumed that my reader is as ignorant and confused as I was when I began. However, as only the ability to add and subtract is needed, together with some very elementary geometry, I am sure that he will not find the subject too complicated. The difficulties begin, as in all navigation, in estimating the weight to put on the information available and to judge correctly the accuracy that has been obtained. If the navigator does not know if his position line is one mile out or twenty it is as if he had never taken a sight. The reader's work then will start after he has read this book, and when he begins to observe often enough to form his own opinion of the value of his sights.

I have avoided going into unnecessary details so that there are a number of slightly inaccurate statements. For instance, I say that one nautical mile equals one minute of arc on the Earth's surface: this is not strictly true, for the measurement varies at the poles and the equator, but this difference is never apparent in practical navigation. Also, for simplicity, I have left out any mention of such technicalities as the celestial equator or the celestial horizon. If the reader wishes to study the theory of navigation more thoroughly, I advise *Foundations of Astronomy* by Smart, *The Principles of Navigation* by Anderson or the *Admiralty Manual of Navigation*, Volume 2, each in its own way a masterpiece of lucidity.

The last twenty years have seen a considerable change in almanacs and tables. The *Air Almanac,* used for example in past editions, is now more specialised and less suitable for yachtsmen while the *Nautical Almanac* has been reorganised and simplified so that I have had, rather unwillingly, to change from one to the other for the examples in this edition. I have, however, kept to the air tables (A.P. 3270) because the present marine ones are unnecessarily complicated and they will be replaced very soon by new marine tables which are about to be published.

Few amateur navigators can hope to reach the standards of a professional, especially of a hydrographer; regular daily observations over the years ensure that confidence and accuracy which mark the expert. But this does not mean that the beginner cannot take adequately accurate and valuable sights, even in rough seas, and I can assure him that the sense of triumph when a sight proves correct is well worth the effort involved. It is the first step which is difficult: take one sight and you will feel bewildered, take two and the fog begins to clear, take a dozen and you will wonder what all the fuss was about.

I have aimed at simplicity throughout this book and can only hope that my explanations are clear enough to encourage a would-be navigator to 'go and take a sight' with an adequate idea of what he is trying to do and a modicum of confidence that he will be able to do it.

Mary Blewitt

1971

ACKNOWLEDGEMENTS

The tables (Appendices A–E) from *The Nautical Almanac*, are Crown copyright, and are reproduced by permission of Her Majesty's Stationery Office. The tables (Appendices F–H) from H.O. 249 (A.P. 3270) are reproduced by permission of the U.S. Naval Oceanographic Office.

CONTENTS

PART ONE: THEORY   I

1 The Heavenly Bodies (1); Geographical Position (1);   I
   Declination (1); Hour Angle (2); Zenith (6); Horizon (6);
   Altitude (7); Zenith Distance (7); Elevated Pole (8);
   Azimuth and Azimuth Angle (8); Great Circles (9);
   Greenwich Mean Time (10)

2 The Position Line (10); The Meridian Passage (13)   10

3 The Spherical Triangle   15

4 Sextant Altitude and True Altitude   19

PART TWO: PRACTICE   21

1 Almanacs; Tables   21

2 Sun Sights (22); Moon Sights (31); Planet Sights (32);   22
   Meridian Sights (34); Star Sights (35); Pole Star Sights (40)

NOTES   46

   Sextants   46

   Star Globes   47

   Spherical Triangles   48

   Tables of Computed Altitude and Azimuth   48
      (H.D. 486, H.O. 214)

   Marine Sight Reduction Tables (N.P. 401, H.O. 229)   49

   Practice Sights   50

   Plotting Sheets   50

APPENDICES   53

INDEX   65

KEY TO THE DIAGRAMS

The key applies to all the diagrams in this book.

P, P′	North and South Poles.
E, E′	Equator.
H, H′	Horizon.
Q	Centre of the Earth.
X	Geographical position of the heavenly body under discussion.
Z	The observer.
Z′	Observer's zenith.
G	Any point on the Greenwich meridian.

From the foregoing it follows that—

The line PZ is part of the observer's meridian.

The line PX is part of the meridian of the geographical position of the heavenly body under discussion.

The line PG is part of the Greenwich meridian.

Except where further description is necessary these letters are not explained again in the book.

ABBREVIATIONS

BST	British Summer Time
DR	Dead reckoning.
GHA	Greenwich hour angle.
GMT	Greenwich Mean Time.
GP	Geographical position.
IE	Index error.
LHA	Local hour angle
SD	Semi-diameter.
A.P.	Air Publication (Air Ministry).
H.D.	Hydrographic Department (Admiralty)
H.O.	Hydrographic Office (U.S. Naval Oceanographic Office).

PART ONE: THEORY

I

Before the theory of a sight can be understood there are certain facts about the Earth which must be thoroughly grasped, and certain terms which must be learned.

THE HEAVENLY BODIES

We navigate by means of the Sun, the Moon, the planets and the stars. Forget the Earth spinning round the Sun with the motionless stars infinite distances away, and imagine that the Earth is the centre of the universe and that all the heavenly bodies circle slowly round us, the stars keeping their relative positions while the Sun, Moon and planets change their positions in relation to each other and to the stars. This pre-Copernican outlook comes easily as we watch the heavenly bodies rise and set, and is a help in practical navigation.

GEOGRAPHICAL POSITION (GP)

At any moment of the day or night there is some spot on the Earth's surface which is directly underneath the Sun. This is the Sun's GP, and it lies where a line drawn from the centre of the Earth to the Sun cuts the Earth's surface. It is shown in Fig. 1 at X. Not only the Sun but all heavenly bodies have GPs, and these positions can be found from the Almanac at any given moment. The GP is measured by declination and hour angle.

DECLINATION (Fig. 2)

The declination of a heavenly body is the latitude of its GP, and is measured exactly as latitude, in degrees north or south of the equator.

Theory

The declination of the Sun moves from 23° N in midsummer when it reaches the tropic of Cancer, to 23° S in midwinter at the tropic of Capricorn; in the spring and autumn, at the equinoxes, the declination is 0° as the Sun crosses the equator. The declination of the Sun alters at an average rate of one degree every four days throughout the year, but there is a considerable variation in the rate of change: at 17h on 10 June 1971 the declination of the Sun will be 23° 00′·1 N, and the

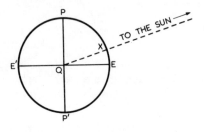

Fig. 1

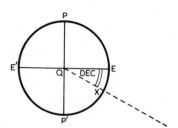

Fig. 2. Angle XQE is the declination of X; it is about 23° S.

GP of the Sun will remain just north of 23° until 10h on 3 July when it will again be 23° 00′·1 (a period of over three weeks, which gives us our long summer evenings). On the other hand, at the equinoxes, the declination changes at a rate of one degree every two and a half days.

The Moon's declination varies between 28° 30′ N and S; it changes rapidly, sometimes as much as six or seven degrees in twenty-four hours. The declinations of the planets change slowly and always lie in a band between 29° N and S. The declinations of the stars are virtually fixed, varying by less than a minute of arc during the year. The declinations of the heavenly bodies are found in the Almanac for every hour of the day and can be interpolated where necessary, with the help of tables, to the nearest minute.

HOUR ANGLE

The GP of any heavenly body is not only on a parallel of latitude but also on a meridian of longitude, and hour angle is the method of measuring this meridian. It differs from longitude in some marked respects.

Let us consider the Sun. If you are standing on the Greenwich meridian, in England, at noon, the Sun is due south of you and its hour angle is nil. Two hours after noon its hour angle is two hours; and as the

Sun sets, goes round the other side of the Earth and rises again, the hour angle increases until at eleven in the morning the hour angle is twenty-three hours, while at noon it comes the full circle of twenty-four hours to start again at nought as it crosses the meridian. The hour angle, when it is measured from the Greenwich meridian, is called the Greenwich Hour Angle (GHA). GHA is always measured in a westerly direction and can be measured in time or in arc, i.e., degrees, minutes and seconds (once round the Earth is 24 hours or 360°).

You might well think that with an accurate watch you could tell how far round the Sun had gone and could measure hour angle just by looking at the time, but this is not so because the Sun does not keep regular, or mean, time, and is sometimes as much as twenty minutes

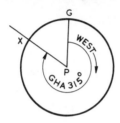

Fig. 3

slow or fast by GMT. This means that the GHA of the Sun has to be looked up in the Almanac where it is tabulated for every hour of every day and, with the help of tables, is then interpolated to the nearest second.

In Fig. 3 we are looking down on to the North Pole. The GHA of the Sun is measured west from the Greenwich meridian, as shown by the arrow. It is morning, for the Sun is coming up to Greenwich and the GHA is approximately 21 hours or 315° (360° = 24 hours, so 15° = 1 hour).

Now hour angle can be measured not only from the Greenwich meridian but from any meridian. When it is measured from the meridian on which you, the observer, are standing it is known as Local Hour Angle (LHA). This is also measured in a westerly direction. If you are west of Greenwich, LHA is less than GHA because the Sun passed Greenwich before it passed you, and so GHA is the larger angle. If you are east of Greenwich, LHA is greater than GHA since the Sun passed you first. Whereas GHA is found from the Almanac, LHA is found by adding or subtracting your longitude to or from GHA.

(At this point you may well ask how you can add or subtract your longitude when it is precisely that which you are trying to determine.

3

Theory

It is a fair question, but just for the moment accept—pretend—that you do know it and later on you will see why you can make this assumption.)

Consider the following four examples:

EXAMPLE A (Fig. 4)

You are somewhere in Canada (Long. 75° W) at 13h local time. As it is one hour after your noon the Sun will be an hour past your meridian and the LHA (heavy line) will be 1 hour (15°). But it is a long time since

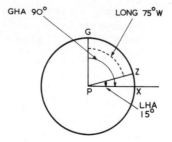

Fig. 4

the Sun crossed the Greenwich meridian; so the GHA (unbroken line) will be much larger. It will be the 75° of your longitude (broken line) plus the 15° the Sun has gone past you, i.e., 90° (6 hours). In west longitudes:

$$\text{GHA} - \text{observer's longitude} = \text{LHA}$$
$$90° - 75° = 15°$$

EXAMPLE B (Fig. 5)

You are in Italy (Long. 15° E) at 11h local time. The LHA (heavy line) is 23 hours or 345°, since it is an hour before your noon. The Sun has

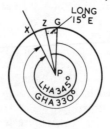

Fig. 5

further to go, however, to reach the Greenwich meridian so GHA (unbroken line) is only 22 hours or 330°. In east longitudes:

4

GHA + observer's longitude = LHA
330° + 15° = 345°

EXAMPLE C (Fig. 6)

You are somewhere in the Atlantic and the Sun has just passed the Greenwich meridian so that GHA is only, say, 1 hour 30 minutes or

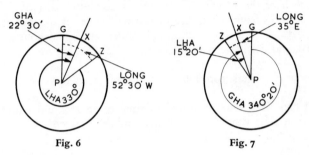

Fig. 6   **Fig. 7**

22° 30′; but the Sun has not yet reached your meridian PZ (Long. 52° 30′ W). On that meridian it is only 10h and the LHA will be 22 hours or 330°. In spite of this the rule holds good, the observer is west of Greenwich so:

GHA — observer's longitude = LHA

but because the subtraction is impossible as it stands (22° 30′ — 52° 30′) we must add 360° to the GHA to get:

382° 30′ — 52° 30′ = 330°

EXAMPLE D (Fig. 7)

Here you are in the eastern Mediterranean (Long. 35° E). The Sun has passed your meridian but has not yet reached Greenwich. LHA therefore is small but GHA is large, say, 340° 20′ or 22 hours 41 minutes. Now your longitude is east of Greenwich so:

GHA + observer's longitude = LHA
340° 20′ + 35°   = 375° 20′

Here we see that the figure is more than 360° so that figure must be subtracted making LHA 15° 20′. We can now say that:

$$LHA = GHA \; {+\ east \atop -\ west} \; longitude$$

If, with an east longitude the total comes to more than 360° that sum (or even occasionally a multiple of that sum) is subtracted to arrive at

5

LHA. With a westerly longitude, if GHA is less than the longitude then 360° is added to GHA to make the subtraction possible.

These diagrams are true not only for the Sun but for all heavenly bodies, but of course only the Sun crosses the Greenwich meridian at noon. The Almanac gives the daily times of the meridian passages at Greenwich of the Moon and planets; it also gives that of the first point of Aries from which the meridian passages of the stars can be calculated (see page 40). A thorough understanding of hour angle is essential, and it will be as well to summarise what has been explained so far.

Hour angle differs from longitude in three main ways:

1. It can always be measured in time or in arc, and conversion tables are given in every almanac and most tables.
2. It is *always* measured in a westerly direction.
3. It may be:
   (a) GHA—measured from the Greenwich meridian;
   (b) LHA—measured from the meridian of the observer;
   (c) Sidereal Hour Angle (SHA). This is explained later on page 35.

The GHA of any heavenly body can be found from the Almanac for any given moment, and LHA is obtained by adding or subtracting the observer's longitude to or from the GHA.

To return for a moment to the GP it should now be clear that the GP of any heavenly body is determined by declination and GHA, and that at any given moment this GP could be plotted on a map, although in fact it is never necessary to do so.

ZENITH

If a line were drawn from the centre of the Earth through you and out into space it would lead to your zenith. In other words, it is the point in space immediately above your head. For instance, if you were standing on the GP of the Sun then the Sun would be in your zenith.

HORIZON

As it is impossible to see round a corner we cannot see much of the surface of the Earth which bends away from us in all directions. The horizon lies in a plane, which at sea level is at a tangent to the Earth's

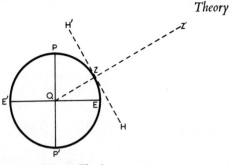

Fig. 8

surface, and that plane is at right angles to the direction of the observer's zenith. In Fig. 8 HH' is a tangent to the Earth's surface at Z and both Z'ZH and Z'ZH' are right angles.

ALTITUDE

The altitude is the angle made at the observer between the Sun (or any other heavenly body) and the horizon directly below it. In Fig. 9 angle

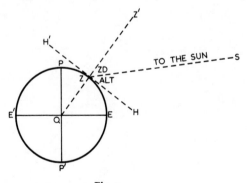

Fig. 9

HZS is the altitude of the Sun. This is the angle you measure with a sextant when you take a sight.

ZENITH DISTANCE

The zenith distance is the complement of the altitude. In Fig. 9 it is angle Z'ZS. Altitude plus zenith distance always equals 90°.

7

ELEVATED POLE

The pole nearer to the observer is called the Elevated Pole: the North Pole in the northern hemisphere and the South Pole in the southern.

AZIMUTH AND AZIMUTH ANGLE

The azimuth is the bearing (true, *not* magnetic) of any heavenly body; this bearing may be called azimuth (Zn) or azimuth angle (Z) depending on the method of measurement. Azimuth angles are measured eastwards or westwards from north or south according to the Elevated Pole: in the northern hemisphere from N to 179° E and from N to 179° W; in the southern hemisphere from south through east or west to 180°. Fig. 10 shows a number of azimuth angles. When working out

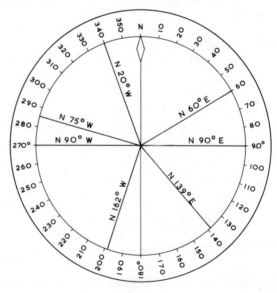

Fig. 10

a sight, azimuth is obtained from the tables where a single figure is found; whether this is N 145° E or N 145° W depends on whether the heavenly body has passed your meridian or not. For example, in the morning the azimuth angle of the Sun will be N and E, in the afternoon N and W. To convert the azimuth angle into azimuth (the bearing

8

measured from north through east from 0° to 360°) the rule in northern latitudes is as follows:

LHA greater than 180° . . . . .Z = Zn
LHA less than 180°. . . . . .360 − Z = Zn

This rule is given on each page of the tables together with the rule for the southern hemisphere. To take an example, in Lat. N 50° with LHA 22° (afternoon with Sun to the west) and Z 145°, then 360° − 145° = 215° = Zn. Had you looked along your compass when you took your sight (allowing for magnetic variation) the Sun would have been on a bearing of 215° from you. It is never possible, however, to measure azimuth accurately enough with a compass and it must be taken from the tables.

GREAT CIRCLES

A great circle is any circle with its centre the centre of the Earth and its radius the distance from the centre to the surface of the Earth. The equator and the meridians are great circles, but the parallels of latitude,

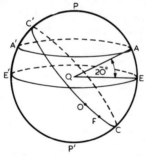

Fig. 11

except the equator, are not, because the centre of a circle formed by a parallel of latitude lies either north or south of the centre of the Earth. Distances along great circles can be measured in two ways, in miles or by the angle subtended at the Earth's centre.

In Fig. 11 three great circles are shown: EE′, CC′ and the circle PEP′E′; AA′ is a parallel of latitude but not a great circle. The distance AE can be measured in arc; it makes an angle of 20° at Q; it can also be measured in miles, and since one minute of arc on the Earth's surface equals one nautical mile, AE = 20 × 60 = 1200 nautical miles. CC′ is a great circle (although it is not a meridian, for it does not pass through the poles), and therefore the distance between O and F, two points on that circle, could also be measured in miles or in arc, being the angle subtended at the centre of the Earth, Q.

This method of measurement can be used between any two points on the Earth's surface (the shortest distance between any two points being part of a great circle) and the interchangeability of arc and mileage should always be kept in mind.

GREENWICH MEAN TIME

GMT is the time given to the world by the Greenwich Observatory. It is an average, or mean, of the Sun's time because the Sun rarely crosses the Greenwich meridian at noon GMT, being erratically fast or slow throughout the year. The Almanac gives the *Equation of Time* daily for midnight and noon and this shows by how much the Sun differs from GMT. The time of the meridian passage alongside tells us if it is fast or slow. For example 17 July 1971 (Appendix A) the equation of time for 12h is 06m 01s, the meridian passage (to the nearest minute) 12h 06m. This tells us that the Sun is 6m 1s slow by GMT.

Before going on to the next part of the book I advise readers to whom all this is new to re-read what has been written so far, because it is important to understand it thoroughly before continuing.

2

THE POSITION LINE

The final result obtained from any sight on any heavenly body is a straight line on your chart, and you are somewhere on that line. If the Sun is in your zenith its altitude is 90°, and there is only one spot on the Earth's surface where you can be—at the GP of the Sun. As you move away from the GP the altitude will lessen, and it will lessen equally whether you go north, south, east or west. However far you move you are on a 'position circle' with its centre at the GP from every point of which the altitude is the same. Fig. 12 shows how the rays from the Sun, or any other heavenly body, strike the Earth; the GP is at X and we see that the further the observer is from X the larger the 'position circle' and the lower the altitude. Altitude lessens until the Sun disappears below the horizon, altitude 0°, zenith distance 90°.

Fig. 13 shows a 'position circle' and an azimuth from the observer to the GP; the Sun is to the SW of the observer who will therefore be on the NE portion of the circle, but unfortunately it is impossible to obtain the azimuth of the Sun accurately enough to fix the exact position on the circle. The only thing to do is to draw a line at right

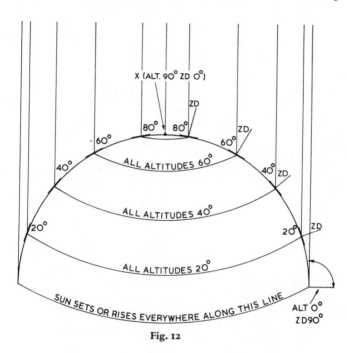

Fig. 12

angles to the most accurate azimuth you can get and say: 'I am somewhere on this line'. The line is drawn straight because the distance from the centre of the circle is so great that it is impossible to show the curve of the circle on the chart.

It may be helpful to realise how very large these 'position circles' are. For example, on a winter morning when the Sun is over SW Africa its altitude in England is about 12° and it has the same altitude near the following places: Greenwich, the Caspian Sea, Madras, the South Pole, Chile, British Guiana and the Azores. Even in midsummer at noon, when the Sun is at its highest (63°) and nearest to us, the circle runs through Greenwich, Istanbul, Cairo, the Congo, the Cape Verde Islands and the Azores.

We have seen that the position line is at right angles to the azimuth

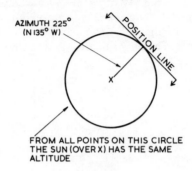

Fig. 13

and this is of practical value, not only for working out sights but also for determining the best time to take them. In Fig. 14 you are approaching a strange coast from the NW and are not certain of your position.

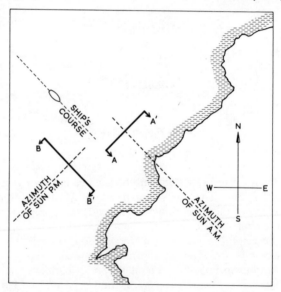

Fig. 14

A morning sight of the Sun when it is SE might give you the line AA' which will determine your distance off land; while from a sight in the afternoon you have the line BB' which positions you along the coast.

THE MERIDIAN PASSAGE

The moment when a heavenly body crosses the meridian of the observer, either to the north or south of him, offers an excellent opportunity for sights for two reasons. First, there is no plotting required; a position line, as we have seen, is at right angles to the bearing of a heavenly body, so when a body crosses your meridian, that is when it is due north or south of you, your position line will run east and west, and a line running east and west is a parallel of latitude. Secondly, accurate time is not required because the moment of the passage is when the altitude of the body is at its highest. Such observations are most commonly used for the Sun, a 'noon sight', and for *Polaris*, the Pole Star, which is permanently on everyone's meridian to the north. There are four cases, each similar in principle.

1. In Fig. 15 PZXP' is the meridian on which both Z and X lie. All the heavenly bodies are so far from the Earth that the rays from any one of them strike the Earth in parallel lines, as a beam of light and not

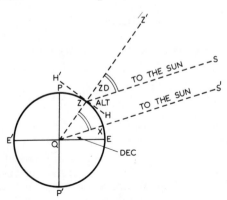

Fig. 15

as a cone (see also Fig. 12). We have here then two parallel lines SZ and S'XQ crossing the straight line Z'ZQ. Therefore angles Z'ZS and ZQX are equal. Now angle Z'ZS is the zenith distance, so when we observe the Sun as it crosses our meridian we learn the size of these angles (ZD = 90° − Alt.). From the Almanac we can find the angle XQE for it is the declination. Add angles ZQX and XQE and the resulting angle ZQE is the latitude of Z. That is to say: latitude = zenith distance + declination.

2. Fig. 16 shows a noon sight with a southerly declination, and it will be seen that here angle ZQX includes angle EQX and so the latitude of Z = ZQX − EQX, that is: latitude = zenith distance − declination.

When declination and latitude are both north or both south this is known as *same* name; when one is north and the other south it is called

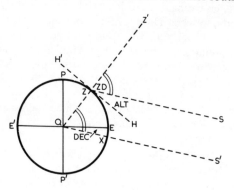

Fig. 16

contrary name. We can therefore say, in cases 1 and 2 where the observer is between the body observed and the elevated pole, that:

$$\text{Lat.} = \text{ZD} \begin{array}{l} + \text{ (same name)} \\ - \text{ (contrary name)} \end{array} \text{Dec.}$$

3. The third case is where the GP of the body observed is between the observer and the Elevated Pole, then: latitude = declination − zenith distance. Observations of *Polaris* present a special case of this. If *Polaris* were exactly over the North Pole (it is not and corrections are necessary) then its GP would always be the North Pole. In Fig. 17 P marks both the GP of *Polaris* and the North Pole because they are the same point and B and B′ mark the direction of the star. Now angles BZZ′ and PQZ are equal and each forms part, respectively, of right angles HZZ′ and PQE, so angles HZB and ZQE are equal. Angle HZB is the altitude of *Polaris* while angle ZQE is the observer's latitude, so the altitude of *Polaris* equals the observer's latitude. Since the declination of *Polaris* is 90° it is equally true to say that:

$$\text{Lat.} = \text{Dec.} - \text{ZD}$$

4. There is one further case (Fig. 18), which mostly occurs in far

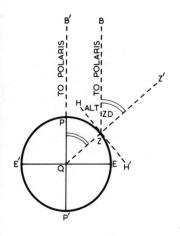

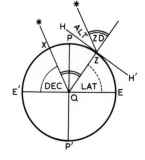

Fig. 17   Fig. 18

northern or southern latitudes, when the observer and the GP of the body observed are on opposite sides of the elevated pole, then:

$$\text{Lat.} = 180° - (\text{Dec.} + \text{ZD})$$

3

THE SPHERICAL TRIANGLE

We have seen that in each of the last four diagrams the zenith distance is equal to the angle subtended at the Earth's centre by the observer and the GP. This is true not only of noon sights or observations of *Polaris*, when the GP is on your meridian, but of all sights, whatever the bearing of the heavenly body. The zenith distance *always* equals the angle at the Earth's centre made by the GP and the observer. Looking again at Figs. 15–18 angle ZQX (ZQP in Fig. 17) will always equal the zenith distance even if the line ZX is not part of a meridian but part of some other great circle. This angle can, of course, be translated into miles at the Earth's surface: an altitude of 47° gives a zenith distance of 43° which, since 1′ of arc = 1 nautical mile, means that the GP is 43 × 60 = 2580 miles away from the observer.

We now know that by taking a sight and finding the zenith distance we can find our distance from the GP of any visible heavenly body, but, because the distances involved are so great, we cannot put a compass on the GP and draw the required circle. Nor can we mark the position line on our chart except when the body is on our meridian and the distances are conveniently marked by parallels of latitude.

We must therefore approach the problem from a different direction. We pretend that we *do* know where we are but that we do *not* know the altitude of the heavenly body. We assume we know our latitude and longitude, that is we work from an *assumed position* (this is sometimes called the chosen position).

In Fig. 19 we are looking at the outside of the Earth. PAP′ and PBP′ are the meridians of X and Z, crossing the equator at A and B. What

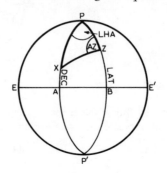

Fig. 19

do we know about the triangle PZX? We know the length of the side PZ; BZ is the observer's latitude, therefore

$$PZ = 90° - \text{latitude of observer.}$$

We know the length of the side PX; AX is the declination, therefore

$$PX = 90° - \text{declination of heavenly body observed.}$$

And we know the included angle ZPX since this is the local hour angle (the angle between the observer's meridian and the meridian of the GP measured in a westerly direction).

Now if we know two sides and the included angle the triangle can be solved, and by spherical trigonometry (or tables) we can find out the length of the side XZ and the other two angles. The length of the side XZ is equal to the zenith distance, so that $90° - XZ = $ altitude. Since this figure is found in the tables it is called the *tabulated altitude* (sometimes called calculated altitude). The tabulated altitude is the altitude we should have got from our sextant if we had taken a sight at our assumed position at that particular time.

Suppose that we take a sight of the Sun and get a true altitude of

16

41° 38′ from the sextant. We know we are somewhere south west of the Isle of Wight and we assume that our position is 50° 00′ N 01° 54′ W (the reasons why we choose this position rather than another will become apparent later). After the necessary calculations we get a tabulated altitude from the tables of 41° 43′; that is to say, if we had been at the above position our true altitude also would have been 41° 43′, but it was not, so we are somewhere else. The difference between the tabulated altitude and the true altitude is 5′, and since 5′ = 5 nautical miles our position line will be 5 miles away from the assumed position. This is called an *intercept* of 5 miles.

If you look again at Fig. 19 you will see that angle PZX is the azimuth, and so much is known about the triangle that this angle can be worked out. In the sight we have just taken the tables tell us that the

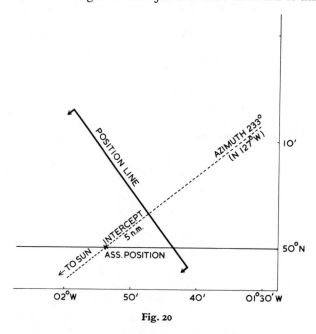

Fig. 20

azimuth angle was 127°, and because it was afternoon we know this was N 127° W (azimuth 233°). On the chart (Fig. 20) we draw the azimuth through the assumed position. As has already been explained, the position line lies at right angles to the azimuth, the intercept was 5 miles, so our position line will be 5 miles *away* from or *towards* the Sun from our assumed position. The further away we are from the GP the less the altitude, so if our true altitude was less than our tabulated altitude we must have been further away than we assumed. In this case the tabulated altitude was 41° 43′, the true altitude 41° 38′, so the position line will be *away* from the Sun. It will be noticed that the position lines are drawn with little arrows pointing *towards* the heavenly body observed; the reason for this is discussed on p. 42. Fig. 21 shows the intercept rule schematically.

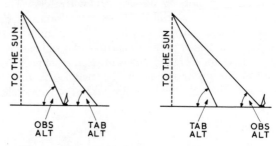

Fig. 21. Observed altitude greater . . . TOWARDS.
Observed altitude less . . . AWAY.

Fig. 22 shows another example of the triangle which must always be solved. In this diagram X is the GP of the Sun. From merely looking at this triangle we can tell certain things: it is morning at Z, for the Sun is to the east, and it is winter for the Sun's declination is south of

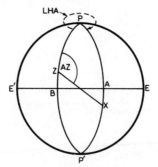

Fig. 22

the equator. Now PZ = 90° − latitude of observer; PX = 90° + declination, and angle ZPX = LHA. In this case LHA, measured west from the meridian of Z, is, say, 323° and the internal angle ZPX is 360° − 323° = 37° but this need not worry us; the important point is that we know the size of the angle. ZX and angle PZX can now be calculated and the tabulated altitude (90° − ZX) and the azimuth angle found in the tables.

This is the principle of every sight taken—that with the known facts about a heavenly body and an assumed position we can calculate an altitude, and that as the tabulated altitude varies from the true altitude so the position line varies from the assumed position.

4

When you take a sight the angle you read off from your sextant is called the sextant altitude (Hs) and there are certain corrections which must be applied to this figure to obtain true altitude (Ho). It is true altitude which is compared with the tabulated altitude (Hc). These abbreviations stand, I believe, for 'height sextant', 'height observed' and 'height calculated' and are used in the tables. The corrections are briefly as follows:

1. *Dip* (*height of eye*). The tabulated altitude in the tables is given as if you were at sea level, but from a boat your eye may be 6 or 60 feet above sea level depending on whether you are in a yacht or a liner. A correction for this must be made to the sextant altitude to make it a 'sea-level' reading; in small boats this is usually about −3′. It must be applied to all sights, except those taken with a bubble sextant or an artificial horizon.

2. *Refraction*. The rays of light from heavenly bodies are bent by the Earth's atmosphere when passing through it, just as a pencil appears bent if you put it in a glass of water. The lower the altitude of the body, the greater is this bending which is called refraction; indeed sights below 6° are inadvisable and those below 10° should be treated with caution. A correction for refraction must be applied to all sights.

3. *Semi-diameter* (Fig. 23). This correction applies only to the Sun and Moon. In theory, when taking a sight the horizon should bisect the 'body', but this is unpractical and when observing the Sun the bottom or *lower* limb is usually rested on the horizon; the *upper* limb

THEORY   LOWER LIMB   UPPER LIMB

Fig. 23

of the Sun can be used if the lower limb is obscured by cloud. With the Moon the upper or lower limb is used, depending on which is available. In each case half the diameter of the body must be allowed for. Semi-diameter is given in the Almanac every three days for both Sun and Moon. The correction, usually about 16′, is added for the lower limb and subtracted for the upper limb. It is not applied if you are using an artificial horizon or a bubble sextant.

4. *Parallax.* It has been mentioned earlier that the rays from heavenly bodies strike the Earth in parallel lines. This is not strictly true, but as far as the Sun, stars and planets are concerned the correction for parallax is negligible and can be ignored. The Moon, however, is so much nearer to the Earth that a considerable correction is usually necessary; it is called horizontal parallax (HP) or parallax in altitude (P in A) and is tabulated in the almanacs.

PART TWO: PRACTICE

I

Two books are needed for working out sights: an almanac and a book of tables. Any almanac can be used with any tables and vice versa even though certain tables and almanacs are designed to be used together.

1. ALMANACS

These are temporary publications, also called ephemerides, which must be renewed every year. They give the positions (GHA and declination) of the heavenly bodies throughout the year, as well as such additional information as times of sunrise and set, eclipses, the Moon's phases, etc. You only need one almanac but there are several to choose from.

The *Nautical Almanac*, used for the examples in this book, is published in one volume for the year; it is designed for marine use and gives the required information in a compact and simple way.

The *Air Almanac*, as its title implies, is designed for use in aircraft. It appears in three volumes each year, with tear-out pages for each day. It is easy to use and if you find it on board you will have no difficulty in adjusting from the *Nautical Almanac*.

Reed's Almanac prints the same information but in a very condensed form requiring much interpolation; amuse yourself trying it out for practice (in case your *Nautical Almanac* goes overboard) but for quick and easy work stick to one of the other two almanacs.

2. TABLES

These are publications which never go out of date, with the one exception of the first volume of A.P. 3270. The simplest and best

21

tables are entered with LHA, declination and latitude to find tabulated altitude and azimuth, that is to say they solve the innumerable variations of the spherical triangle. (For pure interest H.D. 486 has 720 solutions to a full page, 360 pages to a volume and six volumes, making a total of about one and a quarter million solutions.) There are many books of tables but the yacht navigator has three to choose from, although H.D. 486 will probably be out of print soon:

(a) *Sight Reduction Tables for Air Navigation* (A.P. 3270, H.O. 249) are the tables used for the examples in this book. They consist of three volumes, each 31 × 26 cm. Volume 1 will become obsolete in 1975 when a new edition, based on a later epoch, will be issued.

Vol. 1, Selected Stars (Epoch 1970·0)
Vol. 2, Latitudes 0° — 39°, Declinations 0° — 29°
Vol. 3, Latitudes 40° — 89°, Declinations 0° — 29°

(b) *Tables of Computed Altitude and Azimuth* (H.D. 486, H.O. 214) are marine tables which will gradually go out of print now H.O. 229 is published, but the various volumes will be found still in use for many years. There are six volumes, each covering 15° of latitude and declinations from 0° — 75°; they measure 30 × 24 cm. Short notes on their use are given on page 48.

(c) *Sight Reduction Tables for Marine Navigation* (N.P. 401, H.O. 229) are new marine tables. They consist of six volumes, each covering 16° of latitude and declinations from 0° — 90°; they are about the same size as A.P. 3270. These tables are more accurate than the old ones but such accuracy is of no particular advantage in yacht navigation. Notes on the use of H.O. 229 are given on page 49.

2

SUN SIGHTS

Let us pretend that you are in the North Sea, DR position 51° 56′ N, 01° 51′ E, on the morning of 7 December 1971 and that you wish to take a sight of the Sun.

In order to work out your sight you need two things: the sextant altitude of the Sun and the exact time at which you measured that altitude. So far we have looked at the various examples as if we had been able to stop the movement of the heavens and examine the various situations at leisure, but unlike Joshua we cannot make the Sun stand still or stay the Moon; all is in continual movement so that every answer is valid for an infinitesimal period only and we can but try to time it as best we can. How much difference an error in time means varies considerably. In the 'arc-into-time' conversion table we see that $1' = 4s$, so that at the worst an error of 4 seconds could make a difference of 1 n.m.; this means that a watch misread by a minute could cause an error of 15 miles. It is never really as bad as this but a mistake of $1°$ of LHA (4 minutes) can easily put you 45 miles out—so be careful! Do not try at first to take your own time; get someone else to do it for you—and make sure that they can tell the time. It is quite amazing how many people there are who, when you say 'now', do not write down the correct time.

Take your sextant and make yourself comfortable, well supported and firm from the waist down and mobile from the waist up. Make sure that you have a clear view of the Sun and of the horizon below it. Generally, the higher you are the better, because you can more easily avoid false horizons due to wave tops, but it is not worth while making yourself unsteady to gain height. It is also obviously important to choose a place where there is a minimum of spray. Arrange your shade glasses so that you have a good clear Sun which will not dazzle you and use a light horizon glass if it is necessary (usually it is not).

The clarity of the horizon is of the greatest importance. Sometimes in calms there is not enough difference between air and water to distinguish the horizon clearly; on other occasions mist obscures the real horizon and gives a false horizon nearer to you. Very rarely there may be abnormal dip; this can sometimes be noticed because the horizon, seen through the sextant, may appear to 'boil', or the funnels of distant ships seem to reach up into the sky. If you have any doubts about the horizon, if it looks in any way odd, or hazy, either do not take your sights or treat them with caution.

When you have the Sun more or less on the horizon, rock the sextant gently from side to side and you will see the Sun swing as if it

Sun., Tuesday, 7 December 1971

Deck watch			Sextant	
h	m	s	°	′
11	54	31	14	37·5
11	55	56	14	38·0
11	56	37	14	41·5
11	57	34	14	39·0
11	58	12	14	44·0
5)280	170		5)200·0	
11	56	34	14	40·0

were attached to a pendulum; it is at the lowest point of this swing that you take the sight. When you feel sure that the Sun is just resting on the horizon, call to your time-keeper and then read off to him the degrees and minutes from your sextant. Take a series of five sights at about 40-second intervals; they will appear something like the example on page 23. How accurately you read your sights depends in part on the sextant; some read to 0′·1 others to 0′·5. In this book I have taken readings to the nearest half minute which is quite sufficient, at any rate to begin with.

You will notice that in this series of sights the Sun is still rising, indeed it is still an hour before noon, but it is rising very slowly, so much so that a slight error in the observation makes it appear that the Sun is already sinking. With other sights the change in altitude may be much more marked. The object in averaging five sights is to compensate for these little errors. If you can only take three sights, average those but do not rely on the position line from one single observation.

Now with a deck-watch time of 11h 56m 34s and a sextant altitude of 14° 40′ you can start to work out your sight. (See Appendix B.)

1. DECK-WATCH TIME TO GMT

	h	m	s
Deck-watch	11	56	34
D-w corr.			−5
BST	−1	00	00
GMT	10	56	29

The deck-watch is here assumed to be on British Standard Time. This will of course vary in different parts of the world.* BST now no longer applies in winter when clocks are set to GMT.

2. LHA SUN FROM GMT

Each double page in the main body of the *Nautical Almanac* covers three days and gives the GHA and declination for Aries, Sun, Moon and planets, together with the SHA and declination of certain stars and the times of rising and setting of the Sun and Moon. Aries and the stars and planets are found on the left-hand page, the Sun and Moon on the right-hand page. GHA Sun is tabulated every hour and at, say, 10h on 7 December it is 332° 11′.7. This leaves 56m 29s unaccounted for, the figure to be added to GHA for these minutes and seconds being called the *increment*.

At the back of the almanac there are a number of yellow pages headed 'Increments and Corrections'. Appendix C shows the page for 56m and 57m. Three columns ('Sun/Planets', 'Aries' and 'Moon') give the, very similar, figures which must be added to the hourly figure of GHA. In the Sun column against 29s we find 14° 07'·3, which is the increment to be added to GHA Sun for 10h to give us GHA Sun at 10h 56m 29s. The working looks like this:

GHA Sun (10h)	332°	11'·7
Increment (56m 29s)	14	07 ·3
GHA Sun	346	19 ·0

Now the included angle of the spherical triangle was LHA, not GHA, and when discussing hour angles we saw that LHA equals GHA plus or minus longitude; one way to remember this is by the old rhyme:

> Longitude east, GHA least,
> Longitude west, GHA best.

In this case your DR longitude is east so you must add it to get LHA. Personally, I remember the two words 'east add' and write them on every form. One knows, of course, that west requires the opposite sign.

3. THE ASSUMED POSITION

It is necessary at this point to explain the 'assumed position'. In the past the spherical triangle was resolved by the cosine-haversine method and the sight was worked out from the DR position; modern tables, however, cannot print solutions for every minute of arc of latitude, LHA and declination, so the problem must be simplified. The following three rules determine the assumed position:

1. Your assumed position must be as near your DR position as possible.
2. Your assumed latitude must be a whole number of degrees.
3. Your assumed longitude must be so arranged as to make your LHA a whole number of degrees.

Rules 1 and 2 are quite simple and need no explanation; rule 3 is a little more difficult and two imaginary cases are given below:

(a) DR Long. 08° 25' E

GHA	337°	01'
Ass. Long. E	7	59 (E+)
LHA	345	00

(b) DR Long. 04° 50′ W

GHA	27°	32′	
Ass. Long. W	4	32	(E+)
LHA	23	00	

To return to our sight, DR longitude is 01° 51′ E so to get LHA we have:

GHA	346°	19′·0	
Ass. Long. E	1	41 ·0	(E+)
LHA	348	00 ·0	

4. DECLINATION

From the Sun's declination column (Appendix B) take out the declination for 10h (22° 33′·6 S) and note down the value of *d* given at the bottom of the page (0′·3). This figure is the difference in declination during one hour and for the Sun a mean is tabulated for every three days. The *d* correction for the required proportion of the hour is taken out from the '*v* or *d* correction' column on the appropriate page of increments and corrections. In Appendix C we see that in this case the correction is also 0′·3. The sign (+ or −) for *d* is always found by inspection of the declination column; if the declination is increasing the correction must be added, if decreasing subtracted. For example the *d* correction for the Sun will be added from 21 March to 22 June while the declination increases from 0° to 23° N and again from 21 September to 22 December (0° + 23° S), but it must be subtracted from 22 June to 21 September (23° N − 0°) and from 22 December to 21 March (23° S − 0°).

It should be noted that one nautical mile is the maximum error that can arise even if the *d* correction for the Sun is ignored altogether.

5. ASSUMED LATITUDE

As has been said, the assumed latitude must be an integral degree, and since the DR latitude was 51° 56′ N the assumed latitude will be 52° N.

6. SEXTANT ALTITUDE TO TRUE ALTITUDE

As we saw on page 19 the sextant altitude must be corrected. The first correction is for index error, the error of your sextant (see page 47).

In this example an imaginary error of 1′·5 'on' is allowed for. The other corrections are found in the *Nautical Almanac* in the Altitude Correction Tables on the first two pages for the Sun, stars and planets, and at the end for the Moon (Appendices D and E). Under 'Dip', presuming a height of eye above sea level of between 9 and 10 feet, we find a correction of −3′·0, while in the Sun column 'Oct.–Mar., Lower Limb' against 14° 40′ (between 14° 18′ and 14° 42′) we find +12′·6. This last correction includes both refraction and semi-diameter, so we have:

	°	′
Sextant	14	40·0
I.E.		− 1·5
Dip		− 3·0
Alt. Corr.		+12·6
True Alt.	14	48·1

This is the figure that will be compared with the tabulated altitude. You can now put away your Almanac and get out the tables.

Note. The Altitude Correction Tables in the *Nautical Almanac* should theoretically be entered with *apparent altitude* (sextant altitude corrected for index error and dip), but this is unnecessarily precise for anything except possibly the Moon, and sextant altitude can be used instead.

7. TABULATED ALTITUDE AND AZIMUTH

We have now found the three arguments necessary for entering A.P. 3270: *LHA, Declination* and *Assumed Latitude*. In our example (Appendix F):

LHA 348°   Dec. 22° 33′·9 S   Ass. Lat. 52° N

In A.P. 3270, Vol. 3, turn to latitude 52° where there are four headings:

Declination	0°–14°	*same* name as latitude
,,	,,	*contrary* name to latitude
Declination	15°–29°	*same* name as latitude
,,	,,	*contrary* name to latitude

Name refers to 'north' and 'south' and in our example we must look under *contrary* name because declination is south and latitude north.

Each degree of declination has three columns: the tabulated altitude Hc, the difference *d* and the azimuth angle Z (*d* is the difference between the tabulated altitudes of one degree of declination and of the next higher degree and determines what proportion of the minutes of declination are to be added to or subtracted from Hc).

27

Practice

Tabular entry should *always* be for the integral degree of declination numerically *less* than (or equal to) the actual declination. The excess over the integral degree, i.e. the actual minutes of declination itself, are

SUN lower limb

DATE .... 7 December 1971		DR 51° 56' N 01° 51' E	
DECK WATCH ...... 11^h 56^m 34^s		GHA .... 332° 11'·7	
D-W CORR .... − 5		INCREMENT ... 14 07·3	
STANDARD TIME ... −1		GHA SUN ..... 346 19·0	
GMT 10 56 29		ASS. LONG E .. 1 41·0 .....(E+)	
		LHA SUN ..... 348 00·0	
Dec ... 22° 33'·6 .... d $\pm$(1) ... 0'·3			
d ...... + 0·3			
DEC ... 22 33·9 .... N S.		ASS. LAT .... 52° N	

SEXTANT ... 14° 40'·0 ...	Hc ... 15° 16'	d ... −60	z N 169° E
I E .... − 1·5	d .... − 34		
DIP .... − 3·0 ... TAB. ALT. ... 14 42			
ALT.CORR. ... + 12·6 ... TRUE ALT. ... 14 48			
TRUE ALT. ... 14 48·1 ... INTERCEPT .... 6 TOWARDS ~~AWAY~~		Zn ... 169°	

(1) The appropriate sign for d is found by inspection of the Almanac.
If declination is increasing +, if decreasing −.

Fig. 24

referred to as the declination increment. For example, with a declination of 12° 50' the tables will be entered with 12° and the declination increment is 50'.

In our example declination is 22° 34' (to the nearest minute of arc) so 22° is used as argument and 34' is the declination increment. Against LHA 348° we read: Hc 15° 16', d − 60, Z 169°. Now this altitude is

28

correct for 22° declination but must be still corrected for the declination increment. At the back of A.P. 3270 (and on a loose card), you will find 'Table 5—Correction to Tabulated Altitude for Minutes of Declination' (Appendix H). Under *d* 60 and against 34 you find 34 (this will not be the same figure when *d* is not 60). These 34′ must be subtracted as there is a minus sign before the 60 (correspondingly added when there is a plus sign) from 15° 16′ to get the correct tabulated altitude:

$$15° \ 16′ - 34′ = 14° \ 42′$$

The last figure from the tables, under Z, is the azimuth angle and since the Sun has not yet passed your meridian it is N 169° E, an azimuth of 169°.

Now the true altitude was 14° 48′ to the nearest half minute and the tabulated altitude is 14° 42′, so there is an intercept of 6 miles which will be towards the Sun because the true altitude is the greater. This

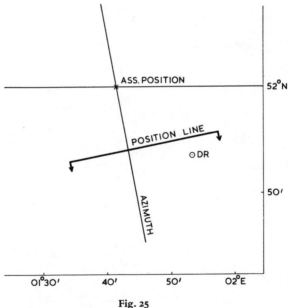

Fig. 25

is clear when you think about it because if the angle is greater, you are more directly underneath the Sun and therefore nearer to it. In practice, however, I advise writing down *true altitude greater towards* where you can always see it and go by rule of thumb.

You are now ready to put the position line on your chart. Mark in the assumed position, draw the azimuth from it (be careful to take it off the true and not the magnetic rose), mark off 6 miles towards the Sun, draw a line at right-angles to the azimuth, and there you are! (Fig. 25.)

In case this seems very complicated Fig. 24 shows the working without explanations—ten small additions or subtractions. Fig. 26 shows another Sun sight taken in July; the pages from the *Nautical Almanac* are reproduced in Appendix A.

SUN lower limb

DATE ..... *15 July 1971* .....   DR .. *49° 50' N 04° 20' W* ....

DECK WATCH   *08ʰ 03ᵐ 22ˢ*   GHA   *283° 33'·0*

D-W CORR.   *-1*   INCREMENT .... *50·3* .....

STANDARD TIME ... *- 1* ....   GHA SUN   *284   23·3*

G M T   *07   03   21*   ASS. LONG *W 4 23·3* ..... (E+)

                                         LHA SUN   *280° 00·0*

Dec .. *21° 37'·3* ..... d ⁺₋ (1)   *0·4* ....

   d .... *00·0* ....

DEC *21   37·3*   N.$\cancel{\text{S}}$.   ASS. LAT.   *50° N*

SEXTANT   *22° 42'·5*   Hc   *22° 15'*   d *+ 44*   z *N 83°E*

I E   ... *- 2·0* ....   d .... *27* ....

DIP   ... *- 3·0* .....   TAB. ALT.   *22 42'* ....

ALT. CORR. *+ 13·7*   TRUE ALT. *22 51'·2*

TRUE ALT. *22 51·2*   INTERCEPT .... *9'·2* ~~TOWARDS~~ ~~AWAY~~   Zn   *83°*

(1) The appropriate sign for d is found by inspection of the Almanac. If declination is increasing +, if decreasing —.

Fig. 26

You will notice that this second sight varies very little from the first: the assumed longitude is subtracted because 'Longitude west, GHA best'; declination and latitude are both north so we look under 'Declination *same* name as Latitude in A.P. 3270; the *d* correction for 59' of declination is plus, but this appears clearly in the tables. Fig. 28 shows this sight plotted together with the Moon sight given as an example below.

MOON SIGHTS

Moon sights are as easy to take as Sun sights and almost as easy to work out; they can also be extremely useful. When the Moon is waning and visible in the morning sky, for instance, a simultaneous Sun and Moon sight (one Sun and one Moon sight taken within a few minutes) gives you two position lines and therefore a fix.

The principle of a Moon sight is exactly the same as that of a Sun sight, but as the Moon moves very irregularly and is also very much closer to the Earth the working is slightly more complicated. In the Moon section of the *Nautical Almanac* (Appendix A) there are five columns which, for every hour, give:

GHA This is taken out for the hour and the appropriate increment for minutes and seconds added, as for the Sun.

v The increment for the Moon is tabulated for the slowest rate of change per hour and because the Moon often moves much faster a correction is usually necessary for this *variation*. On the correct 'Increments and Corrections' page in the '*v* or *d* corr.' columns entering with *v* you will find the figure to be added to GHA Moon.

Dec. This is tabulated every hour as for the Sun.

d The difference is given for every hour for the Moon instead of every three days as for the Sun; the sign must be found by inspection of the declination in just the same way.

H.P. This is the correction for horizontal parallax (see page 20) and the figure is used for entering the second section of the Altitude Correction Tables for the Moon.

Let us look at one example: if a Moon sight were taken with GMT xh 56m 01s and for the day and hour we find the value of *v* to be 6'·2 and that of *d* to be 12'·9, then from 'Increments and Corrections' (Appendix C) we get an increment of 13° 22'·0, *v* correction

31

5'·8 and *d* correction 12'·1. The *v* and *d* correction columns are valid for the entire minute and no account is taken of the seconds of GMT. The *v* correction will always be additive, the sign for *d* will depend on the behaviour of the declination. It will be clear that *v* and *d* for the Moon cannot be ignored.

Usually you can choose by eye which limb of the Moon to observe, but this is impossible when the Moon is full or nearly so.

Sextant altitude to true altitude is also slightly more complicated. Correct the sextant altitude for index error and dip to arrive at apparent altitude and with this figure as argument enter the tables at the back of the *Nautical Almanac* (Appendix E). It is superfluous to discuss here the clear instructions given there—but *do* read them carefully. Here are two examples:

App. Alt.	33° 42'	49° 36'
H.P.	57'·6	61'·5
Limb (U or L)	upper	lower
1st corr.	57'·2	47'·2
2nd corr.	3 ·6	8 ·6
	60 ·8	
Upper limb	−30	
Total corr.	+30 ·8	+55 ·8

Fig. 27 shows a Moon sight taken a few minutes before the last Sun sight; the chart work is in Fig. 28.

In the past the Moon was considered unreliable for observations and one still sometimes hears the same thing said. This was probably due to difficulties with the complicated interpolations required for such a fast-moving body. Modern almanacs make it possible to reduce lunar observations with all the necessary precision. The pleasure and added ease of being able to take sights without the protective dazzle lenses will appeal to all.

PLANET SIGHTS

Planets are extremely useful for navigation as they are clearly visible while there is a good horizon in the morning and evening. The *Nautical Almanac* gives the data for Venus, Mars, Jupiter and Saturn. Values for *v* and *d* are given at the bottom of each column, each figure being a mean for the three days. The only difference from Sun and Moon

MOON upper limb

DATE _15 July 1971_   DR _49° 50' N 04° 20' W_

DECK WATCH _07^h 58^m 35^s_   GHA (6h) _3° 44'·0_   v _10'·4_

D-W CORR _+ 11_   INCR.(58m 46s) _14 01·3_

STANDARD TIME _-1_   v (+) _10·1_   H.P. _59'·2_

GMT _06 58 46_   GHA MOON _17 55·4_

ASS. LONG _W 03 55·4_ (E+)

Dec _13° 09'·2_   d$^+$(1) _14·5_   LHA MOON _14 00·0_

d _14·1_

DEC _13 23·3_ N.   ASS. LAT. _50° N_

SEXTANT _51° 27'·5_   Hc _51° 16'_   d _+58_   z _158°_

I E _- 2·0_   d _22_

DIP _- 3·1_   TAB. ALT. _51 38_

APP. ALT. _51 22·4_   TRUE ALT. _51 42·4_

1ST CORR. _45·9_   INTERCEPT _4·4_ TOWARDS / AWAY   Zn _202°_

2ND CORR. _4·1_

TRUE ALT. _52 12·4_ (LOWER LIMB)

(1) The appropriate sign for Dec. is found by inspection of the Almanac. If Dec. is increasing +, if decreasing −.

3RD CORR. _-30'_ (2)

(2) 3rd correction is for upper limb only.

TRUE ALT. _51 42·4_ (UPPER LIMB)

Fig. 27

sights is that v is sometimes a minus quantity for Venus. The corrections to be applied to sextant altitude are given on pages A2 and A3 in the *Almanac*. Index error and dip must be applied as always and the altitude correction is for refraction only as semi-diameter is not of course required.

33

Planet Identification. The *Nautical Almanac* gives a diagram of the movement of planets with very clear notes on how to use it; it even includes a section headed 'Do not confuse' to help when two planets are close together. If, in spite of this, you find that you are sailing

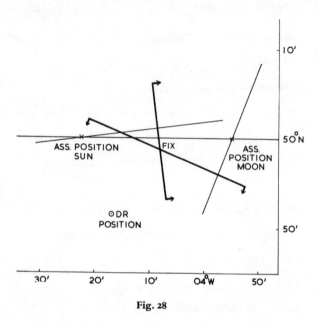

Fig. 28

briskly through Birmingham it is quite probable that you have mistaken your planet or even observed a bright star.

MERIDIAN SIGHTS

A meridian sight of the Sun is usually called a *noon sight* because it is taken at the observer's local noon. When taking a noon sight the first step is to discover the time the Sun will cross your meridian. The time of the Sun's meridian passage at Greenwich is given for each day at the bottom of the page in the *Nautical Almanac*. This time is correct at Greenwich but if you are to the east the passage will be earlier, if to the west later. The degrees of your DR longitude must be multiplied by four to get minutes (of time) which must then be added to or subtracted from the time of the meridian passage at Greenwich to give you the GMT of the meridian passage on your longitude. For example, on 7 December 1971 (Appendix B) the meridian passage of the Sun is at 11h 51m. Now at 10° W the passage will be 40 minutes later (10 × 4), i.e. at 12h 31m GMT; at 10° E the passage will be 40m earlier at 11h 11m.

Let us look at two examples of the working of this sight, the theory of which was explained on page 13, both of the Sun on 15 July 1971 (Appendix A):

	DR Lat. 54° 27′ N			DR Lat. 11° 23′ S	
	Long. 00° 00′			Long. 00° 00′	
	90°	00′·0		90°	00′·0
True Alt.	−57	05 ·6		−57	05 ·6
Z.D.	32	54 ·4		32	54 ·4
Dec. N	+21	35 ·4 (*same* name)		−21	35 ·4 (*contrary* name)
Lat.	54	29 ·8 N		11	19 ·0 S

The Sun remains at the highest point of its trajectory for a number of minutes (even up to eight or ten). When therefore it appears on your sextant not to be rising any longer take a series of five sights spread over two or three minutes, and take the average as your sextant altitude. There is no need to time these sights.

Do not forget that the declination must be taken out for the time of the observation, so that if you are far east or west of Greenwich your noon will be several hours different from that of Greenwich and therefore the appropriate declination must be found.

Meridian sights can be taken for any heavenly body and the times of the meridian passages of the Moon and planets are given in the Almanac at the foot of the page.

Meridian sights have three advantages:

1. Accurate time is not necessary.
2. The working out is very simple.
3. There is no plotting on the chart.

STAR SIGHTS

Sidereal Hour Angle. The GHAs of individual stars are not given in the Almanac. A known point in the heavens, called the First Point of Aries (denoted by the sign ♈) has been chosen, and the GHA of this point is tabulated in the Almanac as if it were a heavenly body. The stars, for navigational purposes, are fixed in relation to each other and to Aries, so that the angle between the meridian of Aries and the meridian of a particular star does not change. This angle, measured in a westerly direction from the meridian of Aries, is the Sidereal Hour

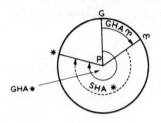

Fig. 29

Angle (SHA). In Fig. 29 we see

GHA♈: the hour angle of the meridian of Aries measured westwards from the Greenwich meridian.

SHA*: the hour angle of the star measured westwards from the meridian of Aries. This angle does not alter.

GHA*: the hour angle of the star measured westwards from the Greenwich meridian.

This last angle *always* equals the sum of the other two hour angles:

$$GHA♈ + SHA* = GHA*$$

Let us look at an example. On 7 December 1971, DR 35° 43′ N 19° 12′ E *Sirius* is observed at 04h 56m 32s:

GHA Aries (04h)	135°	16′·7	Appendix B
Increment (56m 32s)	14	10 ·3	,,　　C
GHA Aries	149	27 ·0	
SHA *Sirius*	259	00 ·9	,,　　B
	408	27 ·9	
	−360		
GHA *Sirius*	48	27 ·9	

As always when dealing with hour angles, if the figure comes to more than 360° that figure is subtracted. The increment to be added to GHA Aries for the minutes and seconds of GMT is found in the Aries column in 'Increments and Corrections'. Having got GHA*, LHA* is obtained in the usual way.

The declinations of stars virtually do not change during the year and are given, together with SHA, in the 'stars' column in the main body of the Almanac. It follows that provided the declination of a star is less than 30° (either north or south) the sight can be worked out from Vols. 2 and 3 of A.P. 3270 as for the other heavenly bodies. Vol. 1, however, is designed for star sights and should normally be used. If you wish to take sights of stars with declinations higher than 30° and which are not in Vol. 1 you must either use other tables, such as H.D. 486, or the cosine-haversine method.

TWILIGHT AND THE PREPARATION OF STAR SIGHTS

The period immediately before sunrise and after sunset is called *civil twilight* and lasts while the Sun is between 0° and 6° below the horizon; the horizon is clear and planets may be observed during this time but the light is too bright for stars. The period when the Sun is between 6° and 12° below the horizon is called *nautical twilight* and is the time for observing stars. Both periods—civil and nautical twilight—are of the same length but the length varies greatly depending on latitude and time of year. The *Nautical Almanac* gives the time when these twilights begin, a mean for three days appearing on each page. The time need not be taken to the nearest minute provided you start early enough, but some preparation is necessary if you are not to struggle up on deck, sextant in hand, to find the horizon merged into the night. It is not easy to pick up a star in the sextant and the approximate altitude should be set on the instrument, then looking along the azimuth the star will appear more or less on the horizon. Let us look at an example of planning, first with individual stars (A.P. 3270, Vols. 2 and 3) and then with Vol. 1 *Selected Stars*. We can work to the nearest degree.

Evening, 15 July 1971, DR 39° N 40° W
Nautical twilight begins at Greenwich   20h 35m GMT
Longitude (40°) × 4   2   40
Nautical twilight begins at Long. 40 W   23   15 GMT

37

Practice

GHA♈ (23h)	278°			*Antares*	*Spica*	*Regulus*
Increment (15m)	4	LHA ♈		242°	242°	242°
	282	SHA *		113	159	208
Long. W	40 (E+)				401	450
					360	360
LHA ♈	242°	LHA *		355	41	90

But look how much easier with Vol. 1! The working is the same as far
as LHA Aries but then, with arguments Lat. 39° N and LHA Aries 242° we
can enter the tables direct and find the altitude and azimuth of each of
the following stars * *Deneb, Altair, Nunki,* * *Antares, Arcturus,* * *Alkaid*
and *Kochab*; these seven are the best for observation at that time and
place. In the tables they are given in order of bearing round the clock
and the three with asterisks are those most suitable for a three-star
fix, the names of bright stars being printed in capitals, fainter ones in
lower case. You are unlikely to observe all of them but five will
suffice to give you a good fix. Remember that each degree of LHA is
four minutes of time so you may note down a later figure of LHA for
each successive star and thus when you take your sights you will still
be able to find them easily.

SELECTED STARS

As we have seen above, Vol. 1 of A.P. 3270 gives seven advantageous
stars to observe for each degree of latitude and of LHA Aries. Instead
of the azimuth angle (Z) azimuth (Zn) is given direct; this is possible
because solutions for both north and south latitudes are tabulated. One
last example (Fig. 30 and Appendix G), in the southern hemisphere,
gives the entire working for a series of seven observations and shows
how simple the calculation is. As far as planning goes each star has been
given 1° of LHA later than the preceding one to allow for the passage
of time. Take at least three sights of each star and average them, then
working them out you notice a certain similarity, almost a rhythm.
Index error, dip, deck-watch error and standard time correction are
the same for all; and we note that as time passes LHA gradually
increases. In every case 360° has to be subtracted from LHA. We see
that by chance *Hamal* is on our meridian and therefore no plotting is
necessary on the chart; the position line will be along latitude 40° 18′ S.

These sights are now plotted on the chart, a long and tedious business
which needs care and accuracy; a degree wrong in plotting an azimuth

38

STARS with H.O.249 (A.P.3270)

DATE ....8 December.... 1971.... DR 40° 23'S 150°08'E Standard Time
+ 10 h

NAUTICAL TWILIGHT BEGINS 20ʰ 52ᵐ   GHA ♈ 226° 30·6

...150... ●LONG x 4ᵐ   - 10 00   INCREMENT ...13 02·1...

NT BEGINS LOCALLY GMT   10 52   GHA ♈ 239 32·7

NT BEGINS SHIP'S TIME   20 52   LONG   E 150 08·0 (E+)

LHA ♈ 29 40·7 ± 360°(WHEN NECESSARY)

STAR & LHA Hc & Zn	Hamal 30 26°40'001	ALDEBARAN 31 23°36'040	RIGEL 32 38°02'065	SIRIUS 33 27°23'089	CANOPUS 34 46°56'129	Peacock 35 35°58'221	FOMALHAUT 36 46°46'227
Hs	26° 26'	23° 58·5	38° 28'	28° 22·5	48° 17'·5	32° 9'	44° 56'·5
I E	+ 2	+ 2	+ 2	+ 2	+ 2	+ 2	+ 2
DIP	– 3	– 3	– 3	– 3	– 3	– 3	– 3
ALT CORR	– 1·9	– 2·2	– 1·2	– 1·9	– 0·8	– 1·5	– 0·9
Ho	26 23·1	23 55·3	38 25·8	28 19·6	48 15·7	32 6·5	44 54·6
D W	20 56 13	21 01 42	21 04 34	21 10 28	21 17 02	21 22 15	21 27 25
DWE & ST	–10 12	10 12	10 12	10 12	10 12	10 12	10 12
G M T	10 56 01	11 01 30	11 04 22	11 10 16	11 16 50	11 22 03	11 27 13
GHA	226°30'·6	241° 33'·1	241° 33'·1	241° 33'·1	241° 33'·1	241° 33'·1	241° 33'·1
INCRE	14 02·6	22·6	1 05·7	2 34·4	4 13·2	5 31·7	6 49·4
GHA	240 33 2	241 55·7	242 38·8	244 07·5	245 46·3	247 04·8	248 22·5
ASS. LONG	150 26·8	150 04·3	150 21·2	149 52·5	150 13·7	149 55·2	150 37·5
L H A	31	32	33	34	36	37	39
Hc	26° 41'	24° 05'	38° 44'	28° 09'	48° 08'	31° 58'	44° 29'
Ho	26 23	23 55·5	38 26	28 19·5	48 15·5	32 6·5	44 54·5
INTERCEPT	18 A	9·5 A	18 A	10·5 T	7·5 T	8·5 T	25·5 T
Zn	000°	039°	064°	089°	128°	220°	265°

Fig. 30

makes a sizable error if there is a big intercept. The most probable position is now chosen and this is discussed on page 41; when using A.P. 3270 Vol. I this position may have to be moved, as described below.

PRECESSION AND NUTATION

Volume I of A.P. 3270 has to be renewed every nine years or so because, entering direct with LHA Aries, no account can be taken of precession and nutation. (Precession: a slow westward motion of the equinoctial points along the ecliptic. Nutation: a fluctuation in the precessional movement of the Earth's pole about the pole of the ecliptic.) A correction for these two is given in Table 5 at the back of Volume I, using as arguments the year, LHA Aries and latitude. In the example here the position is only moved one mile in true bearing 060°, but by 1975 the correction could be as big as five miles. Remember that this correction applies only to a position line or fix deduced from Volume I; if *Polaris* or a planet has been observed together with other stars, their position lines must not be transferred and in this case each star position line should be moved individually before arriving at a position.

MERIDIAN PASSAGE

Any star crosses the observer's meridian when LHA Aries = 360° − SHA*; to find the time of a given LHA Aries the observer's longitude must be added to (W) or subtracted from (E) LHA to get GHA; the time of this GHA is then found from the Almanac. For example to find the time of the meridian passage of *Sirius* at Longitude 10° W on 7 December 1971:

SHA *Sirius* 259° 00′·9   360° − 259° 00′·9 = 100° 59′·1 = LHA
Aries
100° 59′·1 + 10° (W) = 110° 59′·1 =
GHA Aries

From the Almanac we find that at 02h GHA Aries is 105° 11′·8 and that 5° 47′·3 is the increment for 23m 5s, so the meridian passage will be at 02h 23m 5s GMT.

POLE STAR SIGHTS

As has already been said (page 14), if *Polaris* were directly over the North Pole its true altitude would be the latitude of the observer; but *Polaris* can bear up to 2° east or west of north, so a correction is necessary.

When you take a sight of *Polaris*, notice the time—the nearest minute will do—then look up GHA Aries in the Almanac for that time. Let us say your sight was at o6h 33m GMT, 7 December 1971. DR 50° 00′ N 03° 42′ E

GHA (o6h)	165° 21′·6		True Alt.	49° 14′·0
Increment (33m)	8 16 ·4		a_0 (arg. 177° 20′)	1 41 ·8
DR Long. E	3 42	(E+)	a_1 (Lat. 50°)	0 ·6
LHA Aries	177 20 ·0		a_2 (December)	0 ·2
			Sum − 1° = Lat.	49 56 ·6

The three corrections in the right-hand column (a_0, a_1 and a_2) are found in the *Polaris* tables at the end of the Almanac. The sextant altitude is corrected as for a star or planet; index error, dip, and refraction from the Altitude Correction Table for stars and planets on page A2 (this is just a simple refraction table). Although accurate time is not required do not omit to take the average of at least three sextant readings as your sextant altitude.

SELECTING THE POSITION

'The mark of a good navigator is not so much his ability to obtain accurate information as his ability to evaluate and interpret correctly the information available to him.' So says Captain Alton B. Moody,

Fig. 31

U.S.N.R., and we must now make sure that having taken, calculated and plotted our sight, we know how to use it. An experienced observer taking a series of sights in good conditions should finish with a possible position-line error of o′·5 n.m. This means that there is a band, one mile wide, in which the ship lies. Fig. 31 shows how the area of uncertainty from the DR plot is reduced by a single position line. With two position lines the position is defined, given equal confidence in either line, by their point of intersection. With three it is generally taken to be at the centre of the cocked hat, although, for example if there is a systematic error (as from an index or dip error) the position could in fact lie outside it. This is shown in Fig. 32 where a systematic error has moved each position line away from the direction of the bodies observed as shown by the 'azimuth' arrows. The surest way to

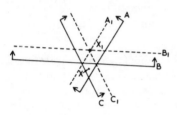

Fig. 32. Three position lines, A,B,C, each with a constant error give the apparent position of the ship at X. $A_1B_1C_1$, the same position lines without the constant error, show the ship to have been in fact at X_1.

Fig. 33. Position-line D enables the constant error to be recognised and allowed for; both sets of position lines, ABCD (with constant error), and $A_1B_1C_1D_1$ (without constant error), show the ship's position to be at X.

evaluate plots is to draw azimuth arrows on each line and to bisect the angle between those which point away from or towards each other; the intersection of these bisectors will then define the position. However, the bisector method is only reliable when the difference in azimuth is greater than 60°. Looking again at Fig. 32 we see that the azimuths were 311°, 000° and 36° there being a 60° difference only between the first and the last. Fig. 33 shows how a fourth position line (with the same systematic error) with an azimuth of 180° clarifies the situation. It should also be noted that the cocked hat in Fig. 32 was not only unreliable because of the bad distribution of azimuth but also suspect because the arrows on lines A and C point in towards the cocked hat while those on B point away; in a truly reliable series of sights the arrows should either all point out from the enclosed area or all point in towards it as in Fig. 33. If some arrows point in, and others out there is an inconsistency which should be resolved before trusting too much to the position obtained.

With several lines, say from six star sights, there will be much more evidence but the plot will also be more difficult to interpret. Fig. 34a shows a real series of six star sights which at first seems most difficult to interpret, there being several cocked hats and enclosed areas. Fig. 34b shows the same sight with azimuth arrows and the bisectors of the

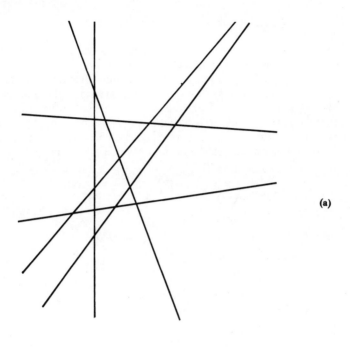

(a)

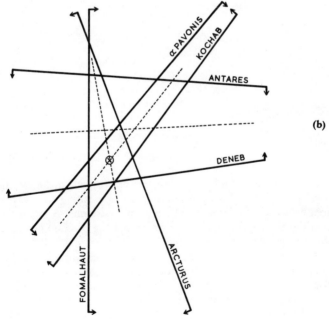

(b)

Fig. 34

43

pairs of position lines most widely separated in azimuth; it is now clear that there was an equal systematic error of five position lines round the most probable position and that only one (*Antares*) has a considerably larger error. Fig. 35 shows our imaginary star sight from page 39 plotted and we can see from the direction of the arrows (all inwards) that the position lines are consistent. It will also be seen that *Rigel* is of little extra use in this case, the bisectors between the three pairs, *Sirius* and *Fomalhaut*, *Aldebaran* and *Peacock*, *Canopus* and *Hamal*, being

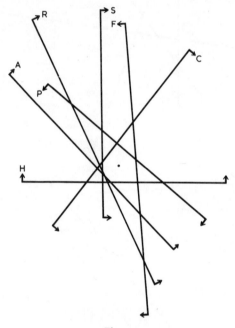

Fig. 35

quite adequate to find a reliable position. In general the position can usually be deduced to sufficient accuracy by taking the centre of the area enclosing the largest number of intersections. As with terrestrial bearings the prudent navigator will adopt the position nearest danger, for example land.

Those who wish to pursue the subject further should read the *Journal of the Institute of Navigation*, Vol. 15, No. 3, page 341, where Michael Richey writes on interpreting astro-position lines at sea and gives the references for a number of articles on the subject, particularly: 'The use of bisectors in selecting the most probable position' by M. Bini in the same Journal, Vol. 8, page 195. Two points remain to be stressed.

First, with modern methods sights are easy to work out. Time, thanks to wireless, presents no serious problems, but *the accuracy of your position line depends on the accuracy of your sight.* You may have a poor horizon, a rough sea, spray coming over the sextant or clouds tending to hide the Sun; any of these will increase your difficulties and no book can help you to overcome them. The only answer is practice and then more practice, until you gain skill and confidence.

Secondly, do not imagine that celestial navigation is only of use if you are crossing the Atlantic; it can be of the greatest value in coastal work. After a night's drifting, for instance, a sight which gives you a position line parallel to your course may enable you to alter it some time before landmarks can be identified, and this alteration may save you several hours sailing against the tide.

Accurate DR is, of course, the basis of good yacht navigation but errors creep in, and to have your DR confirmed or corrected by sights in which you have confidence can save you hours of worry or annoyance. When you can trust your sights you will find that you can use them frequently to good purpose and I hope this book will not only enable you to get your position lines but also encourage you to plunge deeper into the fascinating study of celestial navigation.

NOTES

SEXTANTS

Sextants are so called because the arc at the bottom is one sixth of a circle (60°). The angle it can measure, however, is 120° because moving the index arm 1° moves the image of the Sun 2°. The yachtsman's latest instrument which I have seen (1970) is an octant. It is one eighth of a circle and measures 90°, and is therefore much smaller and easier to stow than a sextant. The principle is exactly the same, however, and we can call them all sextants for convenience.

When you look through the eye-piece of your sextant you see part of a rectangular frame. The left-hand side of this is plain glass through which you can see the horizon, the right-hand side is a mirror; this mirror reflects the light from another mirror at the top of the sextant which is fixed to and swings with the main, or index, arm. The bottom of the index arm swings along a scale which is calibrated in degrees. The minutes are read off the wheel by which the small adjustments are made. In older sextants the minutes are read off the degree scale with the help of a vernier.

Cover the top mirror with one or two of the tinted shades, then face the Sun and look at the horizon (or the garden fence) through the eye-piece. Swing the index arm very slowly until the Sun appears, then with the small adjustor wheel 'move' the Sun until it rests on the horizon; this is your sight. Read your sextant to the nearest half minute; the result is your sextant altitude.

Sextants are delicate instruments and must be treated carefully; various errors can occur of which the chief are as follows:

I. SEXTANT ERROR

This is the basic error of the instrument and should be marked by the manufacturer on the lid of the box. It should be negligible.

2. INDEX ERROR

This is a variable error and must be checked fairly frequently in the following manner: set the sextant roughly at zero and look at the Sun; you will see two suns, bring them so that their edges are just touching and read the sextant. Then reverse the suns and read the sextant again. You will find that one reading will be on the ordinary scale (on) and one on the minus side (off). Subtract the smaller from the greater, halve it, and the result is the index error, to be added if the greater number was 'off', and subtracted if it was 'on'. You can test the accuracy of your reading by adding the two figures; the sum should equal four times the semi-diameter of the Sun. The figures could read like this:

15 July 1971 SD 15'·8

	Sextant readings:	33'·2 on	check:	33'·2
		30'·0 off		30·0
		2)3·2		4)63·2
Index error		1·6 on		15·8

Taken to the nearest half minute the index error is 1'·5 on, and this must be subtracted from the sextant altitude.

Another, but less accurate, way of getting index error if you have a really good hard horizon is to set your sextant at zero, look at the horizon, make it into one straight line and then read the sextant. It should, but will not be zero; the difference is the index error. The Admiralty *Manual of Navigation* advises that sextants should be corrected if the index error is more than 3'.

An imaginary index error is shown in the examples in this book.

3. SIDE ERROR

When you check the index error the two suns should appear exactly one above the other. If they are very much out of alignment the sextant should be corrected.

There are other sextant errors and these, together with methods of correction, can be found in the Admiralty *Manual of Navigation*, Vol. 1.

STAR GLOBES

If you are fortunate enough to own a star globe, or are able to borrow one, study it carefully. It is an easy way to find the position of stars and will also give you a clearer idea of hour angles, azimuths, altitudes and astronomical triangles.

You must imagine that the Earth is inside, at the centre of the globe. The north pole of the globe (the celestial north pole) is over the Earth's North Pole, and the globe's equator (the celestial equator) follows the line of the terrestrial equator. The metal band on the box round the globe is the azimuth ring, and it also marks the celestial horizon, which is the equivalent of your horizon.

Your latitude is set by the fixed metal band passing over the poles and marked in degrees. In north latitudes move the globe's north pole (where *Polaris* is marked) towards north on the azimuth ring until your latitude is opposite the azimuth ring.

The fixed metal band represents your meridian. The globe turning underneath this band shows the stars rising, crossing your meridian and setting. The celestial equator is drawn on the globe itself and is marked both in arc and in time. Let us suppose that you are on the Greenwich meridian at 22 hours on 7 December 1971. Look up GHA Aries for that time which is 46° 01′·1. Find this figure (or rather 46°) on the celestial equator and put it under the fixed band (on the 'south' side of the pole). There will then appear on the globe the stars which would be in the sky at that time. If you are not on the Greenwich meridian you must add or subtract LHA Aries in the usual way.

Now put on the 'brass hat'. The knob at the top is your zenith. From the arms you can read off the altitude and azimuth of any star visible. The astronomical triangle is formed by the pole (P), the knob at the top (Z) and the star (X).

SPHERICAL TRIANGLES

For those interested, the basic formulae for solving the spherical triangle are as follows (Fig. 36):

$$\cos a = \cos b \cos c + \sin b \sin c \cos A$$
$$\cos C = \frac{\cos c - \cos a \cos b}{\sin a \sin b}$$

TABLES OF COMPUTED ALTITUDE AND AZIMUTH (H.D. 486, H.O. 214)

These tables are published in six volumes each of which covers fifteen degrees of latitude for declinations 0° − 75°. Although these marine tables were designed to be used with the *Nautical Almanac* they have not been used for the examples in this book, both because they will shortly be out of date and because they are more complicated to use. They are more accurate, however, and give altitude to o′·1 and

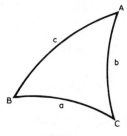

Fig. 36

azimuth to o°·1 (A.P. 3270 gives altitude to 1′ and azimuth to 1°). The main differences from the air tables are as follows:

1. The angle at the elevated pole (P) is used as argument instead of LHA, the rule is

   LHA less than 180° .... P = LHA
   LHA greater than 180° .... P = 360° − LHA

2. The sign for *d* is not given and must be found from the tables by inspection of the tabulated altitude in the next highest column of declination to see if the altitude is increasing or decreasing as the declination increases. If the altitude increases *d* will be plus, if it decreases minus.

3. The altitude is tabulated for every half degree of declination instead of every degree as in the air tables.

MARINE SIGHT REDUCTION TABLES (N.P. 401, H.O. 229)

These new tables are published in six volumes, each covering 16° of latitude; there is an overlap of one degree between volumes. Unlike earlier tables, however, where latitude determines which page to use, here LHA constitutes the main argument. LHA is obtained from the Almanac and the appropriate page found in the tables. On each page latitude is the horizontal argument and declination the vertical. Hc and Z are tabulated to o′·1 and o°·1 respectively, the sign is given for the difference, *d*, the figures for which may be in roman or in italics. The interpolation table for declination increment is more complicated than that in the Air Tables, though naturally, as in all such cases, after

49

a few examples it becomes much simpler through use. If the difference figures are in roman the interpolation is not alarming but if they are in italics, which will not be often as they *can* only occur with altitudes above 60°, a correction for 'double second difference' must be applied which needs a bit of practice. However, these tables will undoubtedly be widely used in the future and I should advise anyone needing celestial navigation seriously, that is for crossing oceans, circling the world, etc., to start off with these tables.

PRACTICE SIGHTS

Useful experience can be gained at home finding out the whereabouts of your house, using an artificial horizon. This is not really a horizon but a horizontal reflecting surface which enables you to measure with a sextant the angle between the heavenly body and its reflection. This angle is double the altitude.

The simplest form of artificial horizon is water in a bucket. This suffers from the disadvantage that it can only be used in calm weather or the ripples distort the reflection. Ideally, a shallow bowl of mercury or oil is used, which gives a better reflection.

On a windless sunny day, place yourself in line with the Sun and the bucket and with the sextant bring the 'real sun' down to its reflection. Superimpose the two suns and take a series of five sights about one a minute, timing them carefully. Average these observations and then halve the resultant angle to obtain sextant altitude. Corrections for dip and semi-diameter are not required but those for index error and refraction must be applied to arrive at true altitude. The Altitude Correction Table for Stars and Planets (Appendix D) is in effect a simple refraction table and should be used. A sight of the Moon is easy to take and is less dazzling than taking a sight of the Sun. The instructions given in the Moon Altitude Correction Tables (Appendix E) for bubble sextants should be followed; also for observations with an artificial horizon.

If your position line does not pass directly through the house remember that, unless you live in Timbuctoo or the wilds of Patagonia, it is you who are wrong and *not* the map!

PLOTTING SHEETS

There are occasions when the Admiralty chart is not suitable for plotting observations. If, for example, you are practising on land with an artificial horizon the chart may not show your area; if far out to sea the chart may not be of a scale to permit accurate plotting of a series of sights, or even if it is the resulting mass of lines might make it almost unusable. Mercator plotting sheets are published and sold by the Admiralty chart department to meet these problems. Each sheet covers 3° of latitude with a scale of 1:670 000. However, it is quite simple to make a plotting sheet for the required latitude, either with a table of secants or geometrically. The ratio longitude:latitude is 1: sec.lat. For example, for latitude 50° (N or S) we find sec. 50° × 1·556;

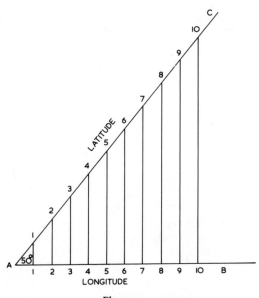

Fig. 37

if therefore the scale of your plotting sheet is to be 1′ longitude = 1 cm, then 1′ latitude will measure 1·556 cm.

Fig. 37 shows how to find this ratio geometrically. Angle A must be the size of the required latitude (here 50°); convenient units of longitude are marked along AB and the perpendiculars drawn to AC. These then mark the corresponding units of latitude. This must be done with great accuracy; five minutes with pencil and paper will convince you that while an error of 1° in low latitudes will cause little harm, any inaccuracy in high latitudes will be reflected directly in the plotted position. Fig. 38 shows a home-made plotting sheet for use with assumed latitude 50° N and any required assumed longitude (here shown between 1° 30′ and 2° 30′ E). On a mercator projection the

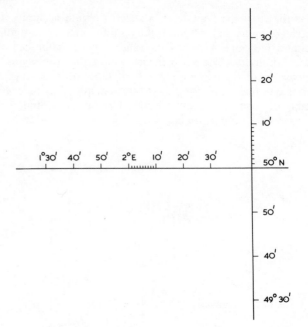

Fig. 38

latitude: longitude ratio changes continuously and home-made sheets should not cover more than one degree of latitude. Using 30′ north and south of a whole degree this allows all the plotting for a given assumed latitude to be done on one sheet. If another degree of latitude is to be used a new sheet must be made. Remember that no reliance should be put on home-made plotting sheets or charts, other than for the limited area described above, as gross errors may result.

APPENDICES

Appendix A

1971 JULY 15, 16, 17 (THURS., FRI., SAT.)

G.M.T.	ARIES G.H.A.	VENUS −3.4 G.H.A.	Dec.	MARS −2.1 G.H.A.	Dec.	JUPITER −1.9 G.H.A.	Dec.	SATURN +0.4 G.H.A.	Dec.	STARS Name	S.H.A.	Dec.
15 00	292 11.7	191 25.7	N23 14.3	325 59.5	S19 38.8	57 32.9	S18 36.9	230 55.9	N18 52.2	Acamar	315 42.2	S 40 24.1
01	307 14.2	206 24.8	14.2	341 02.0	39.0	72 35.4	36.9	245 58.1	52.2	Achernar	335 49.9	S 57 22.4
02	322 16.6	221 24.0	14.0	356 04.6	39.2	87 37.9	36.9	261 00.3	52.3	Acrux	173 45.1	S 62 56.6
03	337 19.1	236 23.1 ..	13.9	11 07.1 ..	39.4	102 40.5 ..	36.9	276 02.5 ..	52.3	Adhara	255 37.7	S 28 55.8
04	352 21.6	251 22.2	13.8	26 09.6	39.6	117 43.0	36.9	291 04.7	52.3	Aldebaran	291 25.8	N 16 27.1
05	7 24.0	266 21.4	13.6	41 12.2	39.8	132 45.6	36.9	306 06.9	52.4			
06	22 26.5	281 20.5	N23 13.5	56 14.7	S19 40.0	147 48.1	S18 36.9	321 09.1	N18 52.4	Alioth	166 48.1	N 56 07.6
07	37 29.0	296 19.6	13.4	71 17.2	40.2	162 50.7	36.9	336 11.3	52.5	Alkaid	153 23.4	N 49 27.1
08	52 31.4	311 18.8	13.2	86 19.8	40.4	177 53.2	36.8	351 13.6	52.5	Al Na'ir	28 22.6	S 47 05.8
09	67 33.9	326 17.9 ..	13.1	101 22.3 ..	40.6	192 55.7 ..	36.8	6 15.8 ..	52.6	Alnilam	276 18.6	S 1 13.0
10	82 36.3	341 17.0	13.0	116 24.8	40.8	207 58.3	36.8	21 18.0	52.6	Alphard	218 27.3	S 8 32.0
11	97 38.8	356 16.2	12.8	131 27.4	41.0	223 00.8	36.8	36 20.2	52.6			
12	112 41.3	11 15.3	N23 12.7	146 29.9	S19 41.2	238 03.4	S18 36.8	51 22.4	N18 52.7	Alphecca	126 37.4	N 26 48.6
13	127 43.7	26 14.4	12.5	161 32.5	41.4	253 05.9	36.8	66 24.6	52.7	Alpheratz	358 16.1	N 28 56.6
14	142 46.2	41 13.6	12.4	176 35.0	41.6	268 08.4	36.8	81 26.8	52.8	Altair	62 38.6	N 8 47.5
15	157 48.7	56 12.7 ..	12.3	191 37.6 ..	41.8	283 11.0 ..	36.8	96 29.0 ..	52.8	Ankaa	353 46.4	S 42 27.3
16	172 51.1	71 11.8	12.1	206 40.1	42.0	298 13.5	36.8	111 31.2	52.9	Antares	113 04.7	S 26 22.4
17	187 53.6	86 11.0	12.0	221 42.6	42.2	313 16.0	36.8	126 33.4	52.9			
18	202 56.1	101 10.1	N23 11.8	236 45.2	S19 42.4	328 18.6	S18 36.8	141 35.6	N18 52.9	Arcturus	146 24.3	N 19 19.8
19	217 58.5	116 09.2	11.7	251 47.7	42.6	343 21.1	36.8	156 37.8	53.0	Atria	108 34.4	S 68 59.0
20	233 01.0	131 08.4	11.5	266 50.3	42.8	358 23.7	36.8	171 40.0	53.0	Avior	234 31.7	S 59 25.3
21	248 03.5	146 07.5 ..	11.4	281 52.9 ..	43.0	13 26.2 ..	36.8	186 42.2 ..	53.1	Bellatrix	279 06.1	N 6 19.7
22	263 05.9	161 06.6	11.2	296 55.4	43.2	28 28.7	36.8	201 44.4	53.1	Betelgeuse	271 35.7	N 7 24.3
23	278 08.4	176 05.8	11.1	311 58.0	43.4	43 31.3	36.7	216 46.7	53.2			
16 00	293 10.8	191 04.9	N23 10.9	327 00.5	S19 43.6	58 33.8	S18 36.7	231 48.9	N18 53.2	Canopus	264 10.7	S 52 40.6
01	308 13.3	206 04.0	10.8	342 03.1	43.8	73 36.3	36.7	246 51.1	53.2	Capella	281 21.4	N 45 58.3
02	323 15.8	221 03.2	10.6	357 05.6	44.0	88 38.9	36.7	261 53.3	53.3	Deneb	49 52.6	N 45 10.6
03	338 18.2	236 02.3 ..	10.4	12 08.2 ..	44.2	103 41.4 ..	36.7	276 55.5 ..	53.3	Denebola	183 05.8	N 14 43.5
04	353 20.7	251 01.4	10.3	27 10.8	44.4	118 43.9	36.7	291 57.7	53.4	Diphda	349 27.3	S 18 08.3
05	8 23.2	266 00.6	10.1	42 13.3	44.6	133 46.5	36.7	306 59.9	53.4			
06	23 25.6	280 59.7	N23 09.9	57 15.9	S19 44.8	148 49.0	S18 36.7	322 02.1	N18 53.5	Dubhe	194 30.2	N 61 54.5
07	38 28.1	295 58.9	09.8	72 18.5	45.0	163 51.5	36.7	337 04.3	53.5	Elnath	278 52.7	N 28 35.2
08	53 30.6	310 58.0	09.6	87 21.0	45.3	178 54.1	36.7	352 06.5	53.5	Eltanin	91 00.3	N 51 29.5
09	68 33.0	325 57.1 ..	09.5	102 23.6 ..	45.5	193 56.6 ..	36.7	7 08.7 ..	53.6	Enif	34 17.1	N 9 44.7
10	83 35.5	340 56.3	09.3	117 26.2	45.7	208 59.1	36.7	22 10.9	53.6	Fomalhaut	15 58.3	S 29 46.2
11	98 38.0	355 55.4	09.1	132 28.7	45.9	224 01.7	36.7	37 13.1	53.7			
12	113 40.4	10 54.5	N23 09.0	147 31.3	S19 46.1	239 04.2	S18 36.7	52 15.4	N18 53.7	Gacrux	172 36.5	S 56 57.4
13	128 42.9	25 53.7	08.8	162 33.9	46.3	254 06.7	36.7	67 17.6	53.8	Gienah	176 24.9	S 17 23.2
14	143 45.3	40 52.8	08.6	177 36.5	46.5	269 09.3	36.7	82 19.8	53.8	Hadar	149 32.8	S 60 14.6
15	158 47.8	55 51.9 ..	08.4	192 39.1 ..	46.7	284 11.8 ..	36.7	97 22.0 ..	53.8	Hamal	328 36.4	N 23 19.8
16	173 50.3	70 51.1	08.3	207 41.6	46.9	299 14.3	36.7	112 24.2	53.9	Kaus Aust.	84 25.1	S 34 24.1
17	188 52.7	85 50.2	08.1	222 44.2	47.1	314 16.9	36.7	127 26.4	53.9			
18	203 55.2	100 49.4	N23 07.9	237 46.8	S19 47.4	329 19.4	S18 36.7	142 28.6	N18 54.0	Kochab	137 17.8	N 74 16.5
19	218 57.7	115 48.5	07.7	252 49.4	47.6	344 21.9	36.6	157 30.8	54.0	Markab	14 09.5	N 15 03.3
20	234 00.1	130 47.6	07.6	267 52.0	47.8	359 24.4	36.6	172 33.0	54.1	Menkar	314 48.1	N 3 58.4
21	249 02.6	145 46.8 ..	07.4	282 54.6 ..	48.0	14 27.0 ..	36.6	187 35.2 ..	54.1	Menkent	148 44.8	S 36 14.7
22	264 05.1	160 45.9	07.2	297 57.1	48.2	29 29.5	36.6	202 37.5	54.1	Miaplacidus	221 47.6	S 69 36.3
23	279 07.5	175 45.0	07.1	312 59.7	48.4	44 32.0	36.6	217 39.7	54.2			
17 00	294 10.0	190 44.2	N23 06.8	328 02.3	S19 48.6	59 34.6	S18 36.6	232 41.9	N18 54.2	Mirfak	309 25.8	N 49 45.6
01	309 12.5	205 43.3	06.7	343 04.9	48.8	74 37.1	36.6	247 44.1	54.3	Nunki	76 36.9	S 26 20.3
02	324 14.9	220 42.5	06.5	358 07.5	49.1	89 39.6	36.6	262 46.3	54.3	Peacock	54 07.9	S 56 49.4
03	339 17.4	235 41.6 ..	06.3	13 10.1 ..	49.3	104 42.1 ..	36.6	277 48.5 ..	54.4	Pollux	244 06.5	N 28 05.6
04	354 19.8	250 40.7	06.1	28 12.7	49.5	119 44.7	36.6	292 50.7	54.4	Procyon	245 33.0	N 5 18.3
05	9 22.3	265 39.9	05.9	43 15.3	49.7	134 47.2	36.6	307 52.9	54.4			
06	24 24.8	280 39.0	N23 05.7	58 17.9	S19 49.9	149 49.7	S18 36.6	322 55.1	N18 54.5	Rasalhague	96 35.3	N 12 34.7
07	39 27.2	295 38.1	05.5	73 20.5	50.1	164 52.2	36.6	337 57.3	54.5	Regulus	208 17.2	N 12 06.5
08	54 29.7	310 37.3	05.3	88 23.1	50.3	179 54.8	36.6	352 59.6	54.6	Rigel	281 41.8	S 8 13.8
09	69 32.2	325 36.4 ..	05.1	103 25.7 ..	50.6	194 57.3 ..	36.6	8 01.8 ..	54.6	Rigil Kent.	140 34.8	S 60 43.4
10	84 34.6	340 35.6	04.9	118 28.3	50.8	209 59.8	36.6	23 04.0	54.7	Sabik	102 48.4	S 15 41.6
11	99 37.1	355 34.7	04.8	133 30.9	51.0	225 02.3	36.6	38 06.2	54.7			
12	114 39.6	10 33.8	N23 04.6	148 33.5	S19 51.2	240 04.9	S18 36.6	53 08.4	N18 54.8	Schedar	350 16.6	N 56 22.7
13	129 42.0	25 33.0	04.4	163 36.1	51.4	255 07.4	36.6	68 10.6	54.8	Shaula	97 04.3	S 37 05.2
14	144 44.5	40 32.1	04.2	178 38.7	51.6	270 09.9	36.6	83 12.8	54.8	Sirius	259 01.8	S 16 40.4
15	159 46.9	55 31.2 ..	04.0	193 41.3 ..	51.9	285 12.4 ..	36.6	98 15.0 ..	54.9	Spica	159 04.5	S 11 00.5
16	174 49.4	70 30.4	03.8	208 44.0	52.1	300 15.0	36.6	113 17.2	54.9	Suhail	223 16.1	S 43 19.3
17	189 51.9	85 29.5	03.5	223 46.6	52.3	315 17.5	36.6	128 19.5	54.9			
18	204 54.3	100 28.7	N23 03.3	238 49.2	S19 52.5	330 20.0	S18 36.6	143 21.7	N18 55.0	Vega	80 59.9	N 38 45.4
19	219 56.8	115 27.8	03.1	253 51.8	52.7	345 22.5	36.6	158 23.9	55.0	Zuben'ubi	137 40.2	S 15 55.6
20	234 59.3	130 26.9	02.9	268 54.4	53.0	0 25.0	36.6	173 26.1	55.1		S.H.A.	Mer. Pass.
21	250 01.7	145 26.1 ..	02.7	283 57.0 ..	53.2	15 27.6 ..	36.6	188 28.3 ..	55.1	Venus	257 54.1	11 16
22	265 04.2	160 25.2	02.5	298 59.7	53.4	30 30.1	36.6	203 30.5	55.2	Mars	33 49.7	2 12
23	280 06.7	175 24.4	02.3	314 02.3	53.6	45 32.6	36.6	218 32.7	55.2	Jupiter	125 23.0	20 02
Mer. Pass.	4 26.5	v −0.9	d 0.2	v 2.6	d 0.2	v 2.5	d 0.0	v 2.2	d 0.0	Saturn	298 38.0	8 31

These pages (reduced) are reproduced from *The Nautical Almanac*

Appendix A (continued)

1971 JULY 15, 16, 17 (THURS., FRI., SAT.)

M.T.	SUN G.H.A.	Dec.	MOON G.H.A.	v	Dec.	d	H.P.
h	° ′	° ′	° ′	′	° ′	′	′
00	178 33·4	N21 40·0	276 46·8	10·7	N11 40·9	14·9	59·3
01	193 33·4	39·6	291 16·5	10·6	11 55·8	14·9	59·3
02	208 33·3	39·3	305 46·1	10·6	12 10·7	14·7	59·3
03	223 33·2	·· 38·9	320 15·7	10·5	12 25·4	14·7	59·2
04	238 33·2	38·5	334 45·2	10·4	12 40·1	14·6	59·2
05	253 33·1	38·1	349 14·6	10·4	12 54·7	14·5	59·2
06	268 33·0	N21 37·7	3 44·0	10·4	N13 09·2	14·5	59·2
07	283 33·0	37·3	18 13·4	10·2	13 23·7	14·4	59·2
08	298 32·9	37·0	32 42·6	10·3	13 38·1	14·3	59·2
09	313 32·8	·· 36·6	47 11·9	10·2	13 52·4	14·2	59·2
10	328 32·8	36·2	61 41·1	10·1	14 06·6	14·2	59·2
11	343 32·7	35·8	76 10·2	10·1	14 20·8	14·0	59·2
12	358 32·6	N21 35·4	90 39·3	10·0	N14 34·8	14·0	59·1
13	13 32·6	35·0	105 08·3	9·9	14 48·8	13·9	59·1
14	28 32·5	34·6	119 37·2	9·9	15 02·7	13·8	59·1
15	43 32·4	·· 34·2	134 06·1	9·9	15 16·5	13·7	59·1
16	58 32·4	33·9	148 35·0	9·8	15 30·2	13·6	59·1
17	73 32·3	33·5	163 03·8	9·7	15 43·8	13·5	59·1
18	88 32·2	N21 33·1	177 32·5	9·6	N15 57·3	13·5	59·1
19	103 32·2	32·7	192 01·1	9·6	16 10·8	13·5	59·1
20	118 32·1	32·3	206 29·7	9·5	16 24·1	13·3	59·0
21	133 32·0	·· 31·9	220 58·2	9·5	16 37·4	13·1	59·0
22	148 32·0	31·5	235 26·7	9·4	16 50·5	13·0	59·0
23	163 31·9	31·1	249 55·1	9·3	17 03·5	13·0	59·0
00	178 31·9	N21 30·7	264 23·4	9·3	N17 16·5	12·8	59·0
01	193 31·8	30·3	278 51·7	9·2	17 29·3	12·8	59·0
02	208 31·7	29·9	293 19·9	9·1	17 42·1	12·6	59·0
03	223 31·7	·· 29·5	307 48·0	9·1	17 54·7	12·6	58·9
04	238 31·6 -	29·1	322 16·1	9·0	18 07·3	12·4	58·9
05	253 31·5	28·7	336 44·1	9·0	18 19·7	12·3	58·9
06	268 31·5	N21 28·3	351 12·1	8·8	N18 32·0	12·2	58·9
07	283 31·4	27·9	5 39·9	8·8	18 44·2	12·1	58·9
08	298 31·4	27·5	20 07·7	8·8	18 56·3	12·0	58·9
09	313 31·3	·· 27·1	34 35·5	8·6	19 08·3	11·8	58·9
10	328 31·2	26·7	49 03·1	8·6	19 20·1	11·8	58·9
11	343 31·2	26·3	63 30·7	8·6	19 31·9	11·6	58·8
12	358 31·1	N21 25·9	77 58·3	8·4	N19 43·5	11·5	58·8
13	13 31·0	25·5	92 25·7	8·4	19 55·0	11·4	58·8
14	28 31·0	25·1	106 53·1	8·3	20 06·4	11·3	58·8
15	43 30·9	·· 24·7	121 20·4	8·3	20 17·7	11·2	58·8
16	58 30·9	24·3	135 47·7	8·2	20 28·9	11·0	58·8
17	73 30·8	23·9	150 14·9	8·1	20 39·9	10·9	58·8
18	88 30·7	N21 23·5	164 42·0	8·1	N20 50·8	10·8	58·7
19	103 30·7	23·1	179 09·1	7·9	21 01·6	10·6	58·7
20	118 30·6	22·7	193 36·0	7·9	21 12·2	10·6	58·7
21	133 30·6	·· 22·3	208 02·9	7·9	21 22·8	10·3	58·7
22	148 30·5	21·9	222 29·8	7·8	21 33·1	10·3	58·7
23	163 30·4	21·4	236 56·6	7·7	21 43·4	10·1	58·7
00	178 30·4	N21 21·0	251 23·3	7·6	N21 53·5	10·0	58·6
01	193 30·3	20·6	265 49·9	7·6	22 03·5	9·9	58·6
02	208 30·3	20·2	280 16·5	7·5	22 13·4	9·6	58·6
03	223 30·2	·· 19·8	294 43·0	7·4	22 23·1	9·6	58·6
04	238 30·2	19·4	309 09·4	7·4	22 32·7	9·4	58·6
05	253 30·1	19·0	323 35·8	7·3	22 42·1	9·4	58·6
06	268 30·0	N21 18·6	338 02·1	7·3	N22 51·5	9·1	58·5
07	283 30·0	18·1	352 28·4	7·2	23 00·6	9·0	58·5
08	298 29·9	17·7	6 54·6	7·1	23 09·6	8·9	58·5
09	313 29·9	·· 17·3	21 20·7	7·1	23 18·5	8·8	58·5
10	328 29·8	16·9	35 46·8	7·0	23 27·3	8·6	58·5
11	343 29·8	16·5	50 12·8	6·9	23 35·9	8·5	58·5
12	358 29·7	N21 16·1	64 38·7	6·9	N23 44·3	8·3	58·4
13	13 29·6	15·6	79 04·6	6·8	23 52·6	8·1	58·4
14	28 29·6	15·2	93 30·4	6·7	24 00·7	8·0	58·4
15	43 29·5	·· 14·8	107 56·1	6·8	24 08·7	7·9	58·4
16	58 29·5	14·4	122 21·9	6·6	24 16·6	7·7	58·4
17	73 29·4	14·0	136 47·5	6·6	24 24·3	7·5	58·4
18	88 29·4	N21 13·5	151 13·1	6·5	N24 31·8	7·4	58·3
19	103 29·3	13·1	165 38·6	6·5	24 39·2	7·3	58·3
20	118 29·3	12·7	180 04·1	6·4	24 46·5	7·0	58·3
21	133 29·2	·· 12·3	194 29·5	6·4	24 53·5	7·0	58·3
22	148 29·2	11·8	208 54·9	6·4	25 00·5	6·7	58·3
23	163 29·1	11·4	223 20·3	6·2	25 07·2	6·7	58·3
	S.D. 15·8	d 0·4	S.D. 16·1		16·0		15·9

Lat.	Twilight Naut.	Civil	Sun-rise	Moonrise 15	16	17	18
°	h m	h m	h m	h m	h m	h m	h m
N 72	□	□	□	19 59	□	□	□
N 70	□	□	□	20 40	□	□	□
68	□	□	□	21 08	20 15	□	□
66	////	////	01 36	21 29	21 05	□	□
64	////	////	02 16	21 47	21 36	21 18	□
62	////	00 51	02 43	22 01	22 00	22 02	22 13
60	////	01 45	03 04	22 13	22 19	22 31	22 57
N 58	////	02 16	03 21	22 24	22 35	22 54	23 27
56	00 47	02 39	03 35	22 33	22 49	23 12	23 50
54	01 37	02 57	03 47	22 41	23 00	23 28	24 08
52	02 05	03 12	03 58	22 49	23 11	23 41	24 24
50	02 27	03 25	04 08	22 55	23 20	23 53	24 38
45	03 05	03 52	04 28	23 10	23 40	24 18	00 18
N 40	03 32	04 12	04 44	23 22	23 56	24 38	00 38
35	03 53	04 28	04 57	23 32	24 10	00 10	00 54
30	04 10	04 42	05 09	23 41	24 22	00 22	01 09
20	04 36	05 05	05 29	23 57	24 43	00 43	01 33
N 10	04 57	05 24	05 46	24 11	00 11	01 01	01 55
0	05 14	05 40	06 02	24 24	00 24	01 18	02 15
S 10	05 30	05 56	06 18	24 37	00 37	01 35	02 35
20	05 44	06 11	06 35	24 51	00 51	01 54	02 57
30	05 59	06 28	06 54	00 01	01 08	02 15	03 22
35	06 07	06 38	07 05	00 07	01 17	02 27	03 37
40	06 15	06 48	07 18	00 14	01 28	02 42	03 54
45	06 24	07 00	07 33	00 23	01 41	02 59	04 15
S 50	06 34	07 14	07 51	00 33	01 57	03 21	04 41
52	06 38	07 21	08 00	00 37	02 04	03 31	04 54
54	06 43	07 28	08 10	00 43	02 13	03 43	05 09
56	06 48	07 36	08 21	00 48	02 22	03 57	05 26
58	06 54	07 45	08 33	00 55	02 33	04 12	05 47
S 60	07 00	07 54	08 48	01 02	02 45	04 31	06 14

Lat.	Sun-set	Twilight Civil	Naut.	Moonset 15	16	17	18
°	h m	h m	h m	h m	h m	h m	h m
N 72	□	□	□	15 56	□	□	□
N 70	□	□	□	15 17	□	□	□
68	□	□	□	14 51	17 35	□	□
66	22 32	////	////	14 31	16 46	□	□
64	21 54	////	////	14 15	16 16	18 32	□
62	21 27	23 14	////	14 02	15 53	17 48	19 39
60	21 08	22 24	////	13 50	15 35	17 19	18 55
N 58	20 50	21 54	////	13 41	15 20	16 57	18 26
56	20 36	21 31	23 19	13 33	15 07	16 39	18 03
54	20 24	21 13	22 33	13 25	14 56	16 24	17 45
52	20 13	20 58	22 05	13 19	14 46	16 11	17 29
50	20 04	20 45	21 44	13 13	14 37	15 59	17 15
45	19 44	20 20	21 06	13 00	14 18	15 36	16 48
N 40	19 28	19 59	20 39	12 49	14 03	15 17	16 26
35	19 14	19 43	20 19	12 40	13 51	15 01	16 08
30	19 02	19 29	20 02	12 33	13 40	14 47	15 53
20	18 43	19 07	19 35	12 19	13 21	14 24	15 26
N 10	18 26	18 48	19 15	12 08	13 04	14 04	15 04
0	18 10	18 32	18 58	11 57	12 49	13 45	14 43
S 10	17 54	18 16	18 42	11 46	12 34	13 26	14 22
20	17 37	18 01	18 28	11 34	12 18	13 06	13 59
30	17 18	17 44	18 13	11 21	12 00	12 43	13 33
35	17 07	17 34	18 06	11 14	11 49	12 30	13 18
40	16 54	17 24	17 57	11 05	11 37	12 14	13 00
45	16 39	17 12	17 48	10 55	11 22	11 56	12 38
S 50	16 21	16 58	17 38	10 43	11 05	11 33	12 11
52	16 12	16 51	17 34	10 38	10 57	11 22	11 58
54	16 02	16 44	17 29	10 31	10 48	11 10	11 43
56	15 51	16 36	17 24	10 25	10 37	10 56	11 26
58	15 38	16 28	17 18	10 17	10 26	10 40	11 04
S 60	15 24	16 18	17 12	10 09	10 13	10 20	10 37

Day	SUN Eqn. of Time 00h	12h	Mer. Pass.	MOON Mer. Pass. Upper	Lower	Age	Phase
	m s	m s	h m	h m	h m	d	
15	05 46	05 49	12 06	05 45	18 10	23	
16	05 52	05 55	12 06	06 36	19 04	24	
17	05 58	06 01	12 06	07 31	20 00	25	☾

Appendix B

1971 DECEMBER 6, 7, 8 (MON., TUES., WED.)

G.M.T.	ARIES G.H.A.	VENUS −3.4 G.H.A.	Dec.	MARS +0.1 G.H.A.	Dec.	JUPITER −1.3 G.H.A.	Dec.	SATURN −0.2 G.H.A.	Dec.	STARS Name	S.H.A.	Dec.
6 00	74 07.7	154 42.4 S24 34.3		85 47.2 S 5 58.2		178 45.3 S22 19.8		13 33.3 N18 30.8		Acamar	315 41.5	S 40 25.0
01	89 10.2	169 41.4	34.2	100 48.2	57.5	193 47.1	19.8	28 36.0	30.8	Achernar	335 49.3	S 57 22.8
02	104 12.6	184 40.5	34.1	115 49.3	56.8	208 49.0	19.9	43 38.6	30.7	Acrux	173 44.8	S 62 56.4
03	119 15.1	199 39.6 ..	33.9	130 50.4 ..	56.2	223 50.8 ..	19.9	58 41.3 ..	30.7	Adhara	255 36.7	S 28 55.8
04	134 17.6	214 38.6	33.8	145 51.5	55.5	238 52.7	20.0	73 44.0	30.7	Aldebaran	291 24.8	N 16 27.4
05	149 20.0	229 37.7	33.7	160 52.6	54.8	253 54.5	20.1	88 46.7	30.6			
06	164 22.5	244 36.7 S24 33.5		175 53.6 S 5 54.2		268 56.4 S22 20.1		103 49.3 N18 30.6		Alioth	166 48.0	N 56 06.4
07	179 25.0	259 35.8	33.4	190 54.7	53.5	283 58.3	20.2	118 52.0	30.6	Alkaid	153 23.6	N 49 26.9
M 08	194 27.4	274 34.8	33.2	205 55.8	52.8	299 00.1	20.3	133 54.7	30.5	Al Na'ir	28 22.7	S 47 06.0
O 09	209 29.9	289 33.9 ..	33.1	220 56.9 ..	52.1	314 02.0 ..	20.3	148 57.3 ..	30.5	Alnilam	276 17.7 ,S	1 13.0
N 10	224 32.3	304 32.9	33.0	235 57.9	51.5	329 03.8	20.4	164 00.0	30.5	Alphard	218 26.6	S 8 32.1
D 11	239 34.8	319 32.0	32.8	250 59.0	50.8	344 05.7	20.4	179 02.7	30.4			
A 12	254 37.3	334 31.0 S24 32.7		266 00.1 S 5 50.1		359 07.5 S22 20.5		194 05.3 N18 30.4		Alphecca	126 37.7	N 26 48.3
Y 13	269 39.7	349 30.1	32.5	281 01.2	49.4	14 09.4	20.6	209 08.0	30.4	Alpheratz	358 15.8	N 28 56.4
14	284 42.2	4 29.1	32.3	296 02.3	48.8	29 11.3	20.6	224 10.7	30.3	Altair	62 38.9	N 8 47.6
15	299 44.7	19 28.2 ..	32.2	311 03.3 ..	48.1	44 13.1 ..	20.7	239 13.4 ..	30.3	Ankaa	353 46.1	S 42 27.6
16	314 47.1	34 27.2	32.1	326 04.4	47.4	59 15.0	20.7	254 16.0	30.3	Antares	113 04.9	S 26 22.3
17	329 49.6	49 26.3	31.9	341 05.5	46.7	74 16.8	20.8	269 18.7	30.2			
18	344 52.1	64 25.3 S24 31.8		356 06.6 S 5 46.1		89 18.7 S22 20.9		284 21.4 N18 30.2		Arcturus	146 24.4	N 19 19.5
19	359 54.5	79 24.4	31.6	11 07.6	45.4	104 20.5	20.9	299 24.0	30.2	Atria	108 35.4	S 68 58.8
20	14 57.0	94 23.4	31.5	26 08.7	44.7	119 22.4	21.0	314 26.7	30.1	Avior	234 30.5	S 59 24.9
21	29 59.5	109 22.5 ..	31.3	41 09.8 ..	44.0	134 24.2 ..	21.0	329 29.4 ..	30.1	Bellatrix	279 05.1	N 6 19.7
22	45 01.9	124 21.6	31.1	56 10.9	43.4	149 26.1	21.1	344 32.1	30.1	Betelgeuse	271 34.7	N 7 24.3
23	60 04.4	139 20.6	31.0	71 11.9	42.7	164 28.0	21.1	359 34.7	30.0			
7 00	75 06.8	154 19.7 S24 30.8		86 13.0 S 5 42.0		179 29.8 S22 21.2		14 37.4 N18 30.0		Canopus	264 09.5	S 52 40.6
01	90 09.3	169 18.7	30.7	101 14.1	41.4	194 31.7	21.3	29 40.1	30.0	Capella	281 20.1	N 45 58.4
02	105 11.8	184 17.8	30.5	116 15.2	40.7	209 33.5	21.3	44 42.7	29.9	Deneb	49 53.1	N 45 11.0
03	120 14.2	199 16.8 ..	30.3	131 16.2 ..	40.0	224 35.4 ..	21.4	59 45.4 ..	29.9	Denebola	183 05.4	N 14 43.6
04	135 16.7	214 15.9	30.2	146 17.3	39.3	239 37.2	21.4	74 48.1	29.9	Diphda	349 26.9	S 18 08.4
05	150 19.2	229 14.9	30.0	161 18.4	38.7	254 39.1	21.5	89 50.7	29.8			
06	165 21.6	244 14.0 S24 29.8		176 19.5 S 5 38.0		269 40.9 S22 21.6		104 53.4 N18 29.8		Dubhe	194 29.5	N 61 53.8
07	180 24.1	259 13.0	29.7	191 20.5	37.3	284 42.8	21.6	119 56.1	29.8	Elnath	278 51.6	N 28 35.2
T 08	195 26.6	274 12.1	29.5	206 21.6	36.6	299 44.7	21.7	134 58.7	29.7	Eltanin	91 01.0	N 51 29.5
U 09	210 29.0	289 11.2 ..	29.3	221 22.7 ..	36.0	314 46.5 ..	21.7	150 01.4 ..	29.7	Enif	34 17.8	N 9 44.9
E 10	225 31.5	304 10.2	29.1	236 23.8	35.3	329 48.4	21.8	165 04.1	29.7	Fomalhaut	15 58.2	S 29 46.4
S 11	240 34.0	319 09.3	29.0	251 24.8	34.6	344 50.2	21.9	180 06.8	29.6			
D 12	255 36.4	334 08.3 S24 28.8		266 25.9 S 5 33.9		359 52.1 S22 21.9		195 09.4 N18 29.6		Gacrux	172 36.2	S 56 57.2
A 13	270 38.9	349 07.4	28.6	281 27.0	33.3	14 53.9	22.0	210 12.1	29.6	Gienah	176 24.5	S 17 23.1
Y 14	285 41.3	4 06.4	28.4	296 28.1	32.6	29 55.8	22.0	225 14.8	29.5	Hadar	149 32.9	S 60 14.2
15	300 43.8	19 05.5 ..	28.2	311 29.1 ..	31.9	44 57.6 ..	22.1	240 17.4 ..	29.5	Hamal	328 35.8	N 23 20.1
16	315 46.3	34 04.6	28.1	326 30.2	31.2	59 59.5	22.2	255 20.1	29.5	Kaus Aust.	84 25.5	S 34 24.1
17	330 48.7	49 03.6	27.9	341 31.3	30.5	75 01.4	22.2	270 22.8	29.4			
18	345 51.2	64 02.7 S24 27.7		356 32.4 S 5 29.9		90 03.2 S22 22.3		285 25.4 N18 29.4		Kochab	137 19.3	N 74 15.9
19	0 53.7	79 01.7	27.5	11 33.4	29.2	105 05.1	22.3	300 28.1	29.4	Markab	14 09.4	N 15 03.4
20	15 56.1	94 00.8	27.3	26 34.5	28.5	120 06.9	22.4	315 30.8	29.3	Menkar	314 47.4	N 3 59.0
21	30 58.6	108 59.8 ..	27.1	41 35.6 ..	27.8	135 08.8 ..	22.5	330 33.4 ..	29.3	Menkent	148 44.8	S 36 13.9
22	46 01.1	123 58.9	26.9	56 36.6	27.2	150 10.6	22.5	345 36.1	29.3	Miaplacidus	221 46.2	S 69 35.8
23	61 03.5	138 58.0	26.8	71 37.7	26.5	165 12.5	22.6	0 38.8	29.2			
8 00	76 06.0	153 57.0 S24 26.6		86 38.8 S 5 25.8		180 14.3 S22 22.6		15 41.5 N18 29.2		Mirfak	309 24.7	N 49 46.0
01	91 08.4	168 56.1	26.4	101 39.9	25.1	195 16.2	22.7	30 44.1	29.2	Nunki	76 37.2	S 26 20.1
02	106 10.9	183 55.1	26.2	116 40.9	24.5	210 18.1	22.7	45 46.8	29.1	Peacock	54 08.5	S 56 49.8
03	121 13.4	198 54.2 ..	26.0	131 42.0 ..	23.8	225 19.9 ..	22.8	60 49.5 ..	29.1	Pollux	244 05.4	N 28 05.7
04	136 15.8	213 53.3	25.8	146 43.1	23.1	240 21.8	22.9	75 52.1	29.1	Procyon	245 32.0	N 5 17.9
05	151 18.3	228 52.3	25.6	161 44.1	22.4	255 23.6	22.9	90 54.8	29.0			
06	166 20.8	243 51.4 S24 25.4		176 45.2 S 5 21.8		270 25.5 S22 23.0		105 57.5 N18 29.0		Rasalhague	96 35.7	N 12 34.7
W 07	181 23.2	258 50.4	25.2	191 46.3	21.1	285 27.3	23.0	121 00.1	29.0	Regulus	208 16.5	N 12 06.2
E 08	196 25.7	273 49.5	25.0	206 47.4	20.4	300 29.2	23.1	136 02.8	28.9	Rigel	281 41.7	S 8 13.8
D 09	211 28.2	288 48.6 ..	24.8	221 48.4 ..	19.7	315 31.0 ..	23.2	151 05.5 ..	28.9	Rigil Kent.	140 35.1	S 60 43.3
N 10	226 30.6	303 47.6	24.6	236 49.5	19.0	330 32.9	23.2	166 08.1	28.9	Sabik	102 48.7	S 15 41.6
E 11	241 33.1	318 46.7	24.4	251 50.6	18.4	345 34.8	23.3	181 10.8	28.8			
S 12	256 35.6	333 45.8 S24 24.2		266 51.6 S 5 17.7		0 36.6 S22 23.3		196 13.5 N18 28.8		Schedar	350 16.1	N 56 23.4
D 13	271 38.0	348 44.8	23.9	281 52.7	17.0	15 38.5	23.4	211 16.1	28.8	Shaula	97 04.7	S 37 05.2
A 14	286 40.5	3 43.9	23.7	296 53.8	16.3	30 40.3	23.4	226 18.8	28.7	Sirius	259 00.9	S 16 41.0
Y 15	301 42.9	18 42.9 ..	23.5	311 54.8 ..	15.7	45 42.2 ..	23.5	241 21.5 ..	28.7	Spica	159 04.3	S 11 00.9
16	316 45.4	33 42.0	23.3	326 55.9	15.0	60 44.0	23.6	256 24.1	28.7	Suhail	223 15.3	S 43 18.9
17	331 47.9	48 41.1	23.1	341 57.0	14.3	75 45.9	23.6	271 26.8	28.6			
18	346 50.3	63 40.1 S24 22.9		356 58.1 S 5 13.6		90 47.7 S22 23.7		286 29.5 N18 28.6		Vega	81 00.5	N 38 45.4
19	1 52.8	78 39.2	22.7	11 59.1	13.0	105 49.6	23.7	301 32.1	28.6	Zuben'ubi	137 40.3	S 15 55.6
20	16 55.3	93 38.2	22.4	27 00.2	12.3	120 51.5	23.8	316 34.8	28.5		S.H.A.	Mer. Pass.
21	31 57.7	108 37.3 ..	22.2	42 01.3 ..	11.6	135 53.3 ..	23.8	331 37.5 ..	28.5	Venus	79 12.8	13 44
22	47 00.2	123 36.4	22.0	57 02.3	10.9	150 55.2	23.9	346 40.1	28.5	Mars	11 06.2	18 14
23	62 02.7	138 35.4	21.8	72 03.4	10.2	165 57.0	24.0	1 42.8	28.4	Jupiter	104 23.0	12 01
Mer. Pass. 18 56.4		v −0.9 d 0.2		v 1.1 d 0.7		v 1.9 d 0.1		v 2.7 d 0.0		Saturn	299 30.5	22 57

These pages (reduced) are reproduced from *The Nautical Almanac*

Appendix B (continued)

1971 DECEMBER 6, 7, 8 (MON., TUES., WED.)

G.M.T.	SUN G.H.A.	SUN Dec.	MOON G.H.A.	v	MOON Dec.	d	H.P.
d h	° ′	° ′	° ′	′	° ′	′	′
6 00	182 20·6	S22 23·3	310 50·7	9·8	N20 26·2	10·1	57·3
01	197 20·4	23·6	325 19·5	9·9	20 16·1	10·2	57·3
02	212 20·1	23·9	339 48·4	10·1	20 05·9	10·3	57·2
03	227 19·9	·· 24·2	354 17·5	10·2	19 55·6	10·4	57·2
04	242 19·6	24·5	8 46·7	10·3	19 45·2	10·5	57·2
05	257 19·3	24·8	23 16·0	10·4	19 34·7	10·6	57·1
06	272 19·1	S22 25·1	37 45·4	10·5	N19 24·1	10·7	57·1
07	287 18·8	25·5	52 14·9	10·6	19 13·4	10·7	57·0
08	302 18·6	25·8	66 44·5	10·8	19 02·7	10·8	57·0
09	317 18·3	·· 26·1	81 14·3	10·9	18 51·9	10·9	57·0
10	332 18·0	26·4	95 44·2	11·0	18 41·0	11·0	56·9
11	347 17·8	26·7	110 14·2	11·1	18 30·0	11·1	56·9
12	2 17·5	S22 27·0	124 44·3	11·2	N18 18·9	11·1	56·9
13	17 17·3	27·3	139 14·5	11·3	18 07·8	11·2	56·8
14	32 17·0	27·6	153 44·8	11·5	17 56·6	11·3	56·8
15	47 16·7	·· 27·9	168 15·3	11·5	17 45·3	11·4	56·8
16	62 16·5	28·2	182 45·8	11·7	17 33·9	11·4	56·7
17	77 16·2	28·5	197 16·5	11·7	17 22·5	11·5	56·7
18	92 15·9	S22 28·8	211 47·2	11·9	N17 11·0	11·6	56·7
19	107 15·7	29·1	226 18·1	12·0	16 59·4	11·6	56·6
20	122 15·4	29·4	240 49·1	12·1	16 47·8	11·7	56·6
21	137 15·2	·· 29·7	255 20·2	12·1	16 36·1	11·8	56·5
22	152 14·9	30·0	269 51·3	12·3	16 24·3	11·8	56·5
23	167 14·6	30·3	284 22·6	12·4	16 12·5	11·9	56·5
7 00	182 14·4	S22 30·6	298 54·0	12·5	N16 00·6	11·9	56·4
01	197 14·1	30·9	313 25·5	12·6	15 48·7	12·0	56·4
02	212 13·8	31·2	327 57·1	12·7	15 36·7	12·1	56·4
03	227 13·6	·· 31·5	342 28·8	12·9	15 24·6	12·1	56·3
04	242 13·3	31·8	357 00·6	12·9	15 12·5	12·1	56·3
05	257 13·0	32·1	11 32·5	13·0	15 00·4	12·2	56·3
06	272 12·8	S22 32·4	26 04·5	13·1	N14 48·2	12·3	56·2
07	287 12·5	32·7	40 36·6	13·2	14 35·9	12·3	56·2
08	302 12·2	33·0	55 08·8	13·3	14 23·6	12·4	56·2
09	317 12·0	·· 33·3	69 41·1	13·3	14 11·2	12·4	56·1
10	332 11·7	33·6	84 13·4	13·5	13 58·8	12·5	56·1
11	347 11·4	33·9	98 45·9	13·5	13 46·3	12·5	56·1
12	2 11·2	S22 34·1	113 18·4	13·7	N13 33·8	12·6	56·0
13	17 10·9	34·4	127 51·1	13·7	13 21·3	12·6	56·0
14	32 10·6	34·7	142 23·8	13·8	13 08·7	12·6	56·0
15	47 10·4	·· 35·0	156 56·6	13·9	12 56·1	12·7	55·9
16	62 10·1	35·3	171 29·5	14·0	12 43·4	12·7	55·9
17	77 09·8	35·6	186 02·5	14·0	12 30·7	12·8	55·9
18	92 09·6	S22 35·9	200 35·5	14·2	N12 17·9	12·8	55·8
19	107 09·3	36·1	215 08·7	14·2	12 05·1	12·8	55·8
20	122 09·0	36·4	229 41·9	14·3	11 52·3	12·9	55·8
21	137 08·8	·· 36·7	244 15·2	14·4	11 39·4	12·9	55·8
22	152 08·5	37·0	258 48·6	14·5	11 26·5	12·9	55·7
23	167 08·2	37·3	273 22·1	14·5	11 13·6	12·9	55·7
8 00	182 08·0	S22 37·5	287 55·6	14·6	N11 00·7	13·0	55·7
01	197 07·7	37·8	302 29·2	14·7	10 47·7	13·1	55·6
02	212 07·4	38·1	317 02·9	14·7	10 34·6	13·0	55·6
03	227 07·1	·· 38·4	331 36·6	14·9	10 21·6	13·1	55·6
04	242 06·9	38·7	346 10·5	14·9	10 08·5	13·1	55·6
05	257 06·6	39·0	0 44·4	14·9	9 55·4	13·1	55·5
06	272 06·3	S22 39·2	15 18·3	15·1	N 9 42·3	13·2	55·5
07	287 06·1	39·5	29 52·4	15·1	9 29·1	13·2	55·5
08	302 05·8	39·8	44 26·5	15·1	9 15·9	13·2	55·4
09	317 05·5	·· 40·0	59 00·6	15·2	9 02·7	13·2	55·4
10	332 05·2	40·3	73 34·8	15·3	8 49·5	13·3	55·4
11	347 05·0	40·6	88 09·1	15·4	8 36·2	13·3	55·4
12	2 04·7	S22 40·8	102 43·5	15·4	N 8 23·0	13·3	55·3
13	17 04·4	41·1	117 17·9	15·5	8 09·7	13·3	55·3
14	32 04·2	41·4	131 52·4	15·5	7 56·4	13·4	55·3
15	47 03·9	·· 41·6	146 26·9	15·6	7 43·0	13·5	55·3
16	62 03·6	41·9	161 01·5	15·6	7 29·7	13·4	55·2
17	77 03·3	42·2	175 36·1	15·7	7 16·3	13·4	55·2
18	92 03·1	S22 42·4	190 10·8	15·8	N 7 02·9	13·3	55·2
19	107 02·8	42·7	204 45·6	15·8	6 49·6	13·5	55·1
20	122 02·5	43·0	219 20·4	15·8	6 36·1	13·4	55·1
21	137 02·2	·· 43·2	233 55·2	15·9	6 22·7	13·4	55·1
22	152 02·0	43·5	248 30·1	15·9	6 09·3	13·5	55·1
23	167 01·7	43·8	263 05·0	16·0	5 55·8	13·4	55·1
	S.D. 16·3	d 0·3	S.D. 15·5		15·3		15·1

Sunrise / Moonrise

Lat.	Twilight Naut.	Twilight Civil	Sun-rise	Moonrise 6	7	8	9
°	h m	h m	h m	h m	h m	h m	h m
N 72	08 04	10 14	■	16 59	20 02	22 11	24 07
N 70	07 46	09 27	■	18 03	20 24	22 20	24 07
68	07 32	08 57	11 12	18 38	20 40	22 27	24 07
66	07 20	08 34	10 04	19 03	20 53	22 33	24 07
64	07 10	08 16	09 29	19 23	21 04	22 38	24 07
62	07 01	08 02	09 04	19 38	21 13	22 42	24 08
60	06 53	07 49	08 45	19 51	21 21	22 46	24 08
N 58	06 46	07 38	08 29	20 02	21 28	22 50	24 08
56	06 40	07 29	08 15	20 12	21 34	22 53	24 08
54	06 34	07 20	08 03	20 21	21 40	22 55	24 08
52	06 29	07 12	07 52	20 28	21 45	22 58	24 08
50	06 24	07 05	07 43	20 35	21 49	23 00	24 08
45	06 13	06 50	07 24	20 50	21 59	23 05	24 08
N 40	06 04	06 37	07 08	21 01	22 07	23 09	24 08
35	05 55	06 26	06 54	21 12	22 13	23 12	24 09
30	05 46	06 16	06 42	21 20	22 19	23 15	24 09
20	05 31	05 58	06 22	21 36	22 29	23 20	24 09
N 10	05 15	05 41	06 04	21 49	22 38	23 25	24 09
0	04 59	05 25	05 47	22 01	22 47	23 29	24 09
S 10	04 41	05 08	05 31	22 13	22 55	23 33	24 09
20	04 19	04 48	05 12	22 26	23 04	23 38	24 10
30	03 51	04 24	04 51	22 41	23 14	23 43	24 10
35	03 34	04 10	04 39	22 50	23 19	23 46	24 10
40	03 11	03 52	04 25	23 00	23 26	23 49	24 10
45	02 43	03 31	04 07	23 11	23 33	23 53	24 10
S 50	02 01	03 03	03 46	23 25	23 42	23 57	24 11
52	01 36	02 49	03 36	23 31	23 47	23 59	24 11
54	01 01	02 32	03 24	23 38	23 51	24 02	00 02
56	////	02 12	03 11	23 46	23 56	24 04	00 04
58	////	01 46	02 56	23 55	24 02	00 02	00 07
S 60	////	01 08	02 37	24 05	00 05	00 08	00 10

Sunset / Moonset

Lat.	Sun-set	Twilight Civil	Twilight Naut.	Moonset 6	7	8	9
°	h m	h m	h m	h m	h m	h m	h m
N 72	■	13 28	15 38	11 59	13 33	12 54	12 25
N 70	■	14 14	15 55	14 59	13 53	13 09	12 43
68	12 29	14 45	16 10	13 17	12 51	12 33	12 18
66	13 35	15 07	16 22	12 51	12 36	12 25	12 15
64	14 12	15 25	16 32	12 30	12 24	12 18	12 13
62	14 38	15 40	16 41	12 14	12 14	12 13	12 11
60	14 57	15 53	16 49	12 00	12 05	12 07	12 09
N 58	15 14	16 04	16 56	11 48	11 57	12 03	12 08
56	15 27	16 13	17 02	11 37	11 50	11 59	12 06
54	15 39	16 22	17 08	11 28	11 43	11 55	12 05
52	15 50	16 30	17 13	11 20	11 38	11 52	12 04
50	15 59	16 37	17 18	11 12	11 33	11 49	12 03
45	16 19	16 52	17 29	10 56	11 21	11 42	12 01
N 40	16 35	17 05	17 39	10 43	11 12	11 37	11 59
35	16 48	17 16	17 48	10 32	11 04	11 32	11 57
30	17 00	17 26	17 56	10 22	10 57	11 27	11 56
20	17 20	17 44	18 12	10 05	10 45	11 20	11 53
N 10	17 38	18 01	18 27	09 50	10 34	11 13	11 51
0	17 55	18 17	18 43	09 36	10 23	11 07	11 48
S 10	18 12	18 35	19 02	09 21	10 13	11 01	11 46
20	18 30	18 54	19 24	09 06	10 02	10 54	11 44
30	18 51	19 19	19 51	08 48	09 49	10 46	11 41
35	19 04	19 33	20 09	08 38	09 42	10 42	11 39
40	19 18	19 51	20 32	08 26	09 33	10 37	11 38
45	19 35	20 12	21 01	08 12	09 23	10 31	11 35
S 50	19 57	20 40	21 43	07 55	09 11	10 23	11 33
52	20 07	20 55	22 08	07 46	09 05	10 20	11 32
54	20 19	21 12	22 43	07 37	08 59	10 16	11 30
56	20 32	21 32	////	07 27	08 52	10 12	11 29
58	20 48	21 58	////	07 15	08 44	10 08	11 27
S 60	21 07	22 38	////	07 01	08 35	10 03	11 26

SUN / MOON

Day	SUN Eqn. of Time 00h	12h	SUN Mer. Pass.	MOON Mer. Pass. Upper	Lower	Age	Phase
	m s	m s	h m	h m	h m	d	
6	09 23	09 11	11 51	03 24	15 49	18	
7	08 58	08 45	11 51	04 12	16 35	19	
8	08 32	08 19	11 52	04 57	17 18	20	

Appendix C

56^m	SUN PLANETS	ARIES	MOON	v or d	Corrn	v or d	Corrn	v or d	Corrn
s	° ′	° ′	° ′	′	′	′	′	′	′
00	14 00·0	14 02·3	13 21·7	0·0	0·0	6·0	5·7	12·0	11·3
01	14 00·3	14 02·6	13 22·0	0·1	0·1	6·1	5·7	12·1	11·4
02	14 00·5	14 02·8	13 22·2	0·2	0·2	6·2	5·8	12·2	11·5
03	14 00·8	14 03·1	13 22·4	0·3	0·3	6·3	5·9	12·3	11·6
04	14 01·0	14 03·3	13 22·7	0·4	0·4	6·4	6·0	12·4	11·7
05	14 01·3	14 03·6	13 22·9	0·5	0·5	6·5	6·1	12·5	11·8
06	14 01·5	14 03·8	13 23·2	0·6	0·6	6·6	6·2	12·6	11·9
07	14 01·8	14 04·1	13 23·4	0·7	0·7	6·7	6·3	12·7	12·0
08	14·02·0	14 04·3	13 23·6	0·8	0·8	6·8	6·4	12·8	12·1
09	14 02·3	14 04·6	13 23·9	0·9	0·8	6·9	6·5	12·9	12·1
10	14 02·5	14 04·8	13 24·1	1·0	0·9	7·0	6·6	13·0	12·2
11	14 02·8	14 05·1	13 24·4	1·1	1·0	7·1	6·7	13·1	12·3
12	14 03·0	14 05·3	13 24·6	1·2	1·1	7·2	6·8	13·2	12·4
13	14 03·3	14 05·6	13 24·8	1·3	1·2	7·3	6·9	13·3	12·5
14	14 03·5	14 05·8	13 25·1	1·4	1·3	7·4	7·0	13·4	12·6
15	14 03·8	14 06·1	13 25·3	1·5	1·4	7·5	7·1	13·5	12·7
16	14 04·0	14 06·3	13 25·6	1·6	1·5	7·6	7·2	13·6	12·8
17	14 04·3	14 06·6	13 25·8	1·7	1·6	7·7	7·3	13·7	12·9
18	14 04·5	14 06·8	13 26·0	1·8	1·7	7·8	7·3	13·8	13·0
19	14 04·8	14 07·1	13 26·3	1·9	1·8	7·9	7·4	13·9	13·1
20	14 05·0	14 07·3	13 26·5	2·0	1·9	8·0	7·5	14·0	13·2
21	14 05·3	14 07·6	13 26·7	2·1	2·0	8·1	7·6	14·1	13·3
22	14 05·5	14 07·8	13 27·0	2·2	2·1	8·2	7·7	14·2	13·4
23	14 05·8	14 08·1	13 27·2	2·3	2·2	8·3	7·8	14·3	13·5
24	14 06·0	14 08·3	13 27·5	2·4	2·3	8·4	7·9	14·4	13·6
25	14 06·3	14 08·6	13 27·7	2·5	2·4	8·5	8·0	14·5	13·7
26	14 06·5	14 08·8	13 27·9	2·6	2·4	8·6	8·1	14·6	13·7
27	14 06·8	14 09·1	13 28·2	2·7	2·5	8·7	8·2	14·7	13·8
28	14 07·0	14 09·3	13 28·4	2·8	2·6	8·8	8·3	14·8	13·9
29	14 07·3	14 09·6	13 28·7	2·9	2·7	8·9	8·4	14·9	14·0
30	14 07·5	14 09·8	13 28·9	3·0	2·8	9·0	8·5	15·0	14·1
31	14 07·8	14 10·1	13 29·1	3·1	2·9	9·1	8·6	15·1	14·2
32	14 08·0	14 10·3	13 29·4	3·2	3·0	9·2	8·7	15·2	14·3
33	14 08·3	14 10·6	13 29·6	3·3	3·1	9·3	8·8	15·3	14·4
34	14 08·5	14 10·8	13 29·8	3·4	3·2	9·4	8·9	15·4	14·5
35	14 08·8	14 11·1	13 30·1	3·5	3·3	9·5	8·9	15·5	14·6
36	14 09·0	14 11·3	13 30·3	3·6	3·4	9·6	9·0	15·6	14·7
37	14 09·3	14 11·6	13 30·6	3·7	3·5	9·7	9·1	15·7	14·8
38	14 09·5	14 11·8	13 30·8	3·8	3·6	9·8	9·2	15·8	14·9
39	14 09·8	14 12·1	13 31·0	3·9	3·7	9·9	9·3	15·9	15·0
40	14 10·0	14 12·3	13 31·3	4·0	3·8	10·0	9·4	16·0	15·1
41	14 10·3	14 12·6	13 31·5	4·1	3·9	10·1	9·5	16·1	15·2
42	14 10·5	14 12·8	13 31·8	4·2	4·0	10·2	9·6	16·2	15·3
43	14 10·8	14 13·1	13 32·0	4·3	4·0	10·3	9·7	16·3	15·4
44	14 11·0	14 13·3	13 32·2	4·4	4·1	10·4	9·8	16·4	15·4
45	14 11·3	14 13·6	13 32·5	4·5	4·2	10·5	9·9	16·5	15·5
46	14 11·5	14 13·8	13 32·7	4·6	4·3	10·6	10·0	16·6	15·6
47	14 11·8	14 14·1	13 32·9	4·7	4·4	10·7	10·1	16·7	15·7
48	14 12·0	14 14·3	13 33·2	4·8	4·5	10·8	10·2	16·8	15·8
49	14 12·3	14 14·6	13 33·4	4·9	4·6	10·9	10·3	16·9	15·9
50	14 12·5	14 14·8	13 33·7	5·0	4·7	11·0	10·4	17·0	16·0
51	14 12·8	14 15·1	13 33·9	5·1	4·8	11·1	10·5	17·1	16·1
52	14 13·0	14 15·3	13 34·1	5·2	4·9	11·2	10·5	17·2	16·2
53	14 13·3	14 15·6	13 34·4	5·3	5·0	11·3	10·6	17·3	16·3
54	14 13·5	14 15·8	13 34·6	5·4	5·1	11·4	10·7	17·4	16·4
55	14 13·8	14 16·1	13 34·9	5·5	5·2	11·5	10·8	17·5	16·5
56	14 14·0	14 16·3	13 35·1	5·6	5·3	11·6	10·9	17·6	16·6
57	14 14·3	14 16·6	13 35·3	5·7	5·4	11·7	11·0	17·7	16·7
58	14 14·5	14 16·8	13 35·6	5·8	5·5	11·8	11·1	17·8	16·8
59	14 14·8	14 17·1	13 35·8	5·9	5·6	11·9	11·2	17·9	16·9
60	14 15·0	14 17·3	13 36·1	6·0	5·7	12·0	11·3	18·0	17·0

57^m	SUN PLANETS	ARIES	MOON	v or d	Corrn	v or d	Corrn	v or d	Corrn
s	° ′	° ′	° ′	′	′	′	′	′	′
00	14 15·0	14 17·3	13 36·1	0·0	0·0	6·0	5·8	12·0	11·5
01	14 15·3	14 17·6	13 36·3	0·1	0·1	6·1	5·8	12·1	11·6
02	14 15·5	14 17·8	13 36·5	0·2	0·2	6·2	5·9	12·2	11·7
03	14 15·8	14 18·1	13 36·8	0·3	0·3	6·3	6·0	12·3	11·8
04	14 16·0	14 18·3	13 37·0	0·4	0·4	6·4	6·1	12·4	11·9
05	14 16·3	14 18·6	13 37·2	0·5	0·5	6·5	6·2	12·5	12·0
06	14 16·5	14 18·8	13 37·5	0·6	0·6	6·6	6·3	12·6	12·1
07	14 16·8	14 19·1	13 37·7	0·7	0·7	6·7	6·4	12·7	12·2
08	14 17·0	14 19·3	13 38·0	0·8	0·8	6·8	6·5	12·8	12·3
09	14 17·3	14 19·6	13 38·2	0·9	0·9	6·9	6·6	12·9	12·4
10	14 17·5	14 19·8	13 38·4	1·0	1·0	7·0	6·7	13·0	12·5
11	14 17·8	14 20·1	13 38·7	1·1	1·1	7·1	6·8	13·1	12·6
12	14 18·0	14 20·3	13 38·9	1·2	1·2	7·2	6·9	13·2	12·7
13	14 18·3	14 20·6	13 39·2	1·3	1·2	7·3	7·0	13·3	12·8
14	14 18·5	14 20·9	13 39·4	1·4	1·3	7·4	7·1	13·4	12·8
15	14 18·8	14 21·1	13 39·6	1·5	1·4	7·5	7·2	13·5	12·9
16	14 19·0	14 21·4	13 39·9	1·6	1·5	7·6	7·3	13·6	13·0
17	14 19·3	14 21·6	13 40·1	1·7	1·6	7·7	7·4	13·7	13·1
18	14 19·5	14 21·9	13 40·3	1·8	1·7	7·8	7·5	13·8	13·2
19	14 19·8	14 22·1	13 40·6	1·9	1·8	7·9	7·6	13·9	13·3
20	14 20·0	14 22·4	13 40·8	2·0	1·9	8·0	7·7	14·0	13·4
21	14 20·3	14 22·6	13 41·1	2·1	2·0	8·1	7·8	14·1	13·5
22	14 20·5	14 22·9	13 41·3	2·2	2·1	8·2	7·9	14·2	13·6
23	14 20·8	14 23·1	13 41·5	2·3	2·2	8·3	8·0	14·3	13·7
24	14 21·0	14 23·4	13 41·8	2·4	2·3	8·4	8·1	14·4	13·8
25	14 21·3	14 23·6	13 42·0	2·5	2·4	8·5	8·1	14·5	13·9
26	14 21·5	14 23·9	13 42·3	2·6	2·5	8·6	8·2	14·6	14·0
27	14 21·8	14 24·1	13 42·5	2·7	2·6	8·7	8·3	14·7	14·1
28	14 22·0	14 24·4	13 42·7	2·8	2·7	8·8	8·4	14·8	14·2
29	14 22·3	14 24·6	13 43·0	2·9	2·8	8·9	8·5	14·9	14·3
30	14 22·5	14 24·9	13 43·2	3·0	2·9	9·0	8·6	15·0	14·4
31	14 22·8	14 25·1	13 43·4	3·1	3·0	9·1	8·7	15·1	14·5
32	14 23·0	14 25·4	13 43·7	3·2	3·1	9·2	8·8	15·2	14·6
33	14 23·3	14 25·6	13 43·9	3·3	3·2	9·3	8·9	15·3	14·7
34	14 23·5	14 25·9	13 44·2	3·4	3·3	9·4	9·0	15·4	14·8
35	14 23·8	14 26·1	13 44·4	3·5	3·4	9·5	9·1	15·5	14·9
36	14 24·0	14 26·4	13 44·6	3·6	3·5	9·6	9·2	15·6	15·0
37	14 24·3	14 26·6	13 44·9	3·7	3·5	9·7	9·3	15·7	15·1
38	14 24·5	14 26·9	13 45·1	3·8	3·6	9·8	9·4	15·8	15·2
39	14 24·8	14 27·1	13 45·4	3·9	3·7	9·9	9·5	15·9	15·3
40	14 25·0	14 27·4	13 45·6	4·0	3·8	10·0	9·6	16·0	15·4
41	14 25·3	14 27·6	13 45·8	4·1	3·9	10·1	9·7	16·1	15·5
42	14 25·5	14 27·9	13 46·1	4·2	4·0	10·2	9·8	16·2	15·6
43	14 25·8	14 28·1	13 46·3	4·3	4·1	10·3	9·9	16·3	15·7
44	14 26·0	14 28·4	13 46·5	4·4	4·2	10·4	10·0	16·4	15·8
45	14 26·3	14 28·6	13 46·8	4·5	4·3	10·5	10·1	16·5	15·9
46	14 26·5	14 28·9	13 47·0	4·6	4·4	10·6	10·2	16·6	15·9
47	14 26·8	14 29·1	13 47·3	4·7	4·5	10·7	10·3	16·7	16·0
48	14 27·0	14 29·4	13 47·5	4·8	4·6	10·8	10·4	16·8	16·1
49	14 27·3	14 29·6	13 47·7	4·9	4·7	10·9	10·4	16·9	16·2
50	14 27·5	14 29·9	13 48·0	5·0	4·8	11·0	10·5	17·0	16·3
51	14 27·8	14 30·1	13 48·2	5·1	4·9	11·1	10·6	17·1	16·4
52	14 28·0	14 30·4	13 48·5	5·2	5·0	11·2	10·7	17·2	16·5
53	14 28·3	14 30·6	13 48·7	5·3	5·1	11·3	10·8	17·3	16·6
54	14 28·5	14 30·9	13 48·9	5·4	5·2	11·4	10·9	17·4	16·7
55	14 28·8	14 31·1	13 49·2	5·5	5·3	11·5	11·0	17·5	16·8
56	14 29·0	14 31·4	13 49·4	5·6	5·4	11·6	11·1	17·6	16·9
57	14 29·3	14 31·6	13 49·7	5·7	5·5	11·7	11·2	17·7	17·0
58	14 29·5	14 31·9	13 49·9	5·8	5·6	11·8	11·3	17·8	17·1
59	14 29·8	14 32·1	13 50·1	5·9	5·7	11·9	11·4	17·9	17·2
60	14 30·0	14 32·4	13 50·4	6·0	5·8	12·0	11·5	18·0	17·3

This page from *The Nautical Almanac* shows increments and corrections for 56m and 57m

Appendix D

A2 ALTITUDE CORRECTION TABLES 10°–90°—SUN, STARS, PLANETS

OCT.—MAR. SUN APR.—SEPT.						STARS AND PLANETS			DIP					
App. Alt.	Lower Limb	Upper Limb	App. Alt.	Lower Limb	Upper Limb	App. Alt.	Corrⁿ	App. Alt. Additional Corrⁿ	Ht. of Eye	Corrⁿ	Ht. of Eye	Corrⁿ	Ht. of Eye	Corrⁿ
								1971	m		ft.		m	
9 34	+10.8	−21.5	9 39	+10.6	−21.2	9 56	−5.3	**VENUS**	2.4	−2.8	8.0		1.0	− 1.8
9 45	+10.9	−21.4	9 51	+10.7	−21.1	10 08	−5.2	Jan. 1–Jan. 17	2.6	−2.9	8.6		1.5	− 2.2
9 56	+11.0	−21.3	10 03	+10.8	−21.0	10 20	−5.1	46° +0.3	2.8	−3.0	9.2		2.0	− 2.5
10 08	+11.1	−21.2	10 15	+10.9	−20.9	10 33	−5.0		3.0	−3.1	9.8		2.5	− 2.8
10 21	+11.2	−21.1	10 27	+11.0	−20.8	10 46	−4.9	Jan. 18–Mar. 5	3.2	−3.2	10.5		3.0	− 3.0
10 34	+11.3	−21.0	10 40	+11.1	−20.7	11 00	−4.8	47° +0.2	3.4	−3.3	11.2		See table ←	
10 47	+11.4	−20.9	10 54	+11.2	−20.6	11 14	−4.7		3.6	−3.4	11.9			
11 01	+11.5	−20.8	11 08	+11.3	−20.5	11 29	−4.6	Mar. 6–Dec. 31	3.8	−3.5	12.6		m	
11 15	+11.6	−20.7	11 23	+11.4	−20.4	11 45	−4.5	42° +0.1	4.0	−3.6	13.3		20	− 7.9
11 30	+11.7	−20.6	11 38	+11.5	−20.3	12 01	−4.4		4.3	−3.7	14.1		22	− 8.3
11 46	+11.8	−20.5	11 54	+11.6	−20.2	12 18	−4.3	**MARS**	4.5	−3.8	14.9		24	− 8.6
12 02	+11.9	−20.4	12 10	+11.7	−20.1	12 35	−4.2	Jan. 1–Apr. 19	4.7	−3.9	15.7		26	− 9.0
12 19	+12.0	−20.3	12 28	+11.8	−20.0	12 54	−4.1	60° +0.1	5.0	−4.0	16.5		28	− 9.3
12 37	+12.1	−20.2	12 46	+11.9	−19.9	13 13	−4.0		5.2	−4.1	17.4			
12 55	+12.2	−20.1	13 05	+12.0	−19.8	13 33	−3.9	Apr. 20–June 13	5.5	−4.2	18.3		30	− 9.6
13 14	+12.3	−20.0	13 24	+12.1	−19.7	13 54	−3.8	41° +0.2	5.8	−4.3	19.1		32	−10.0
13 35	+12.4	−19.9	13 45	+12.2	−19.6	14 16	−3.7	75° +0.1	6.1	−4.4	20.1		34	−10.3
13 56	+12.5	−19.8	14 07	+12.3	−19.5	14 40	−3.6		6.3	−4.5	21.0		36	−10.6
14 18	+12.6	−19.7	14 30	+12.4	−19.4	15 04	−3.5	June 14–Oct. 12	6.6	−4.6	22.0		38	−10.8
14 42	+12.7	−19.6	14 54	+12.5	−19.3	15 30	−3.4	34° +0.3	6.9	−4.7	22.9			
15 06	+12.8	−19.5	15 19	+12.6	−19.2	15 57	−3.3	60° +0.2	7.2	−4.8	23.9		40	−11.1
15 32	+12.9	−19.4	15 46	+12.7	−19.1	16 26	−3.2	80° +0.1	7.5	−4.9	24.9		42	−11.4
15 59	+13.0	−19.3	16 14	+12.8	−19.0	16 56	−3.1		7.9	−5.0	26.0		44	−11.7
16 28	+13.1	−19.2	16 44	+12.9	−18.9	17 28	−3.0	Oct. 13–Dec. 8	8.2	−5.1	27.1		46	−11.9
16 59	+13.2	−19.1	17 15	+13.0	−18.8	18 02	−2.9	41° +0.2	8.5	−5.2	28.1		48	−12.2
17 32	+13.3	−19.0	17 48	+13.1	−18.7	18 38	−2.8	75° +0.1	8.8	−5.3	29.2		ft.	
18 06	+13.4	−18.9	18 24	+13.2	−18.6	19 17	−2.7		9.2	−5.4	30.4		2	− 1.4
18 42	+13.5	−18.8	19 01	+13.3	−18.5	19 58	−2.6	Dec. 9–Dec. 31	9.5	−5.5	31.5		4	− 1.9
19 21	+13.6	−18.7	19 42	+13.4	−18.4	20 42	−2.5	60° +0.1	9.9	−5.6	32.7		6	− 2.4
20 03	+13.7	−18.6	20 25	+13.5	−18.3	21 28	−2.4		10.3	−5.7	33.9		8	− 2.7
20 48	+13.8	−18.5	21 11	+13.6	−18.2	22 19	−2.3		10.6	−5.8	35.1		10	− 3.1
21 35	+13.9	−18.4	22 00	+13.7	−18.1	23 13	−2.2		11.0	−5.9	36.3		See table ←	
22 26	+14.0	−18.3	22 54	+13.8	−18.0	24 11	−2.1		11.4	−6.0	37.6			
23 22	+14.1	−18.2	23 51	+13.9	−17.9	25 14	−2.0		11.8	−6.1	38.9		ft.	
24 21	+14.2	−18.1	24 53	+14.0	−17.8	26 22	−1.9		12.2	−6.2	40.1		70	− 8.1
25 26	+14.3	−18.0	26 00	+14.1	−17.7	27 36	−1.8		12.6	−6.3	41.5		75	− 8.4
26 36	+14.4	−17.9	27 13	+14.2	−17.6	28 56	−1.7		13.0	−6.4	42.8		80	− 8.7
27 52	+14.5	−17.8	28 33	+14.3	−17.5	30 24	−1.6		13.4	−6.5	44.2		85	− 8.9
29 15	+14.6	−17.7	30 00	+14.4	−17.4	32 00	−1.5		13.8	−6.6	45.5		90	− 9.2
30 46	+14.7	−17.6	31 35	+14.5	−17.3	33 45	−1.4		14.2	−6.7	46.9		95	− 9.5
32 26	+14.8	−17.5	33 20	+14.6	−17.2	35 40	−1.3		14.7	−6.8	48.4			
34 17	+14.9	−17.4	35 17	+14.7	−17.1	37 48	−1.2		15.1	−6.9	49.8		100	− 9.7
36 20	+15.0	−17.3	37 26	+14.8	−17.0	40 08	−1.1		15.5	−7.0	51.3		105	− 9.9
38 36	+15.1	−17.2	39 50	+14.9	−16.9	42 44	−1.0		16.0	−7.1	52.8		110	−10.2
41 08	+15.2	−17.1	42 31	+15.0	−16.8	45 36	−0.9		16.5	−7.2	54.3		115	−10.4
43 59	+15.3	−17.0	45 31	+15.1	−16.7	48 47	−0.8		16.9	−7.3	55.8		120	−10.6
47 10	+15.4	−16.9	48 55	+15.2	−16.6	52 18	−0.7		17.4	−7.4	57.4		125	−10.8
50 46	+15.5	−16.8	52 44	+15.3	−16.5	56 11	−0.6		17.9	−7.5	58.9			
54 49	+15.6	−16.7	57 02	+15.4	−16.4	60 28	−0.5		18.4	−7.6	60.5		130	−11.1
59 23	+15.7	−16.6	61 51	+15.5	−16.3	65 08	−0.4		18.8	−7.7	62.1		135	−11.3
64 30	+15.8	−16.5	67 17	+15.6	−16.2	70 11	−0.3		19.3	−7.8	63.8		140	−11.5
70 12	+15.9	−16.4	73 16	+15.7	−16.1	75 34	−0.2		19.8	−7.9	65.4		145	−11.7
76 26	+16.0	−16.3	79 43	+15.8	−16.0	81 13	−0.1		20.4	−8.0	67.1		150	−11.9
83 05	+16.1	−16.2	86 32	+15.9	−15.9	87 03	0.0		20.9	−8.1	68.8		155	−12.1
90 00			90 00			90 00			21.4		70.5			

App. Alt. = Apparent altitude = Sextant altitude corrected for index error and dip.
For daylight observations of Venus, see page 260.

This page is reproduced from *The Nautical Almanac*

Appendix E

ALTITUDE CORRECTION TABLES 0°–35°—MOON

App. Alt.	0°–4° Corrn	5°–9° Corrn	10°–14° Corrn	15°–19° Corrn	20°–24° Corrn	25°–29° Corrn	30°–34° Corrn	App. Alt.
00	0 33.8	5 58.2	10 62.1	15 62.8	20 62.2	25 60.8	30 58.9	00
10	35.9	58.5	62.2	62.8	62.1	60.8	58.8	10
20	37.8	58.7	62.2	62.8	62.1	60.7	58.8	20
30	39.6	58.9	62.3	62.8	62.1	60.7	58.7	30
40	41.2	59.1	62.3	62.8	62.0	60.6	58.6	40
50	42.6	59.3	62.4	62.7	62.0	60.6	58.5	50
00	1 44.0	6 59.5	11 62.4	16 62.7	21 62.0	26 60.5	31 58.5	00
10	45.2	59.7	62.4	62.7	61.9	60.4	58.4	10
20	46.3	59.9	62.5	62.7	61.9	60.4	58.3	20
30	47.3	60.0	62.5	62.7	61.9	60.3	58.2	30
40	48.3	60.2	62.5	62.7	61.8	60.3	58.2	40
50	49.2	60.3	62.6	62.7	61.8	60.2	58.1	50
00	2 50.0	7 60.5	12 62.6	17 62.7	22 61.7	27 60.1	32 58.0	00
10	50.8	60.6	62.6	62.6	61.7	60.1	57.9	10
20	51.4	60.7	62.6	62.6	61.6	60.0	57.8	20
30	52.1	60.9	62.7	62.6	61.6	59.9	57.8	30
40	52.7	61.0	62.7	62.6	61.5	59.9	57.7	40
50	53.3	61.1	62.7	62.6	61.5	59.8	57.6	50
00	3 53.8	8 61.2	13 62.7	18 62.5	23 61.5	28 59.7	33 57.5	00
10	54.3	61.3	62.7	62.5	61.4	59.7	57.4	10
20	54.8	61.4	62.7	62.5	61.4	59.6	57.4	20
30	55.2	61.5	62.8	62.5	61.3	59.6	57.3	30
40	55.6	61.6	62.8	62.4	61.3	59.5	57.2	40
50	56.0	61.6	62.8	62.4	61.2	59.4	57.1	50
00	4 56.4	9 61.7	14 62.8	19 62.4	24 61.2	29 59.3	34 57.0	00
10	56.7	61.8	62.8	62.3	61.1	59.3	56.9	10
20	57.1	61.9	62.8	62.3	61.1	59.2	56.9	20
30	57.4	61.9	62.8	62.3	61.0	59.1	56.8	30
40	57.7	62.0	62.8	62.2	60.9	59.1	56.7	40
50	57.9	62.1	62.8	62.2	60.9	59.0	56.6	50

H.P.	L U	L U	L U	L U	L U	L U	L U	H.P.
54.0	0.3 0.9	0.3 0.9	0.4 1.0	0.5 1.1	0.6 1.2	0.7 1.3	0.9 1.5	54.0
54.3	0.7 1.1	0.7 1.2	0.7 1.2	0.8 1.3	0.9 1.4	1.1 1.5	1.2 1.7	54.3
54.6	1.1 1.4	1.1 1.4	1.1 1.4	1.2 1.5	1.3 1.6	1.4 1.7	1.5 1.8	54.6
54.9	1.4 1.6	1.5 1.6	1.5 1.6	1.6 1.7	1.6 1.8	1.8 1.9	1.9 2.0	54.9
55.2	1.8 1.8	1.8 1.8	1.9 1.9	1.9 1.9	2.0 2.0	2.1 2.1	2.2 2.2	55.2
55.5	2.2 2.0	2.2 2.0	2.3 2.1	2.3 2.1	2.4 2.2	2.4 2.3	2.5 2.4	55.5
55.8	2.6 2.2	2.6 2.2	2.6 2.3	2.7 2.3	2.7 2.4	2.8 2.4	2.9 2.5	55.8
56.1	3.0 2.4	3.0 2.5	3.0 2.5	3.1 2.6	3.1 2.6	3.2 2.7	3.2 2.7	56.1
56.4	3.4 2.7	3.4 2.7	3.4 2.7	3.4 2.7	3.4 2.8	3.5 2.8	3.5 2.9	56.4
56.7	3.7 2.9	3.7 2.9	3.8 2.9	3.8 2.9	3.8 3.0	3.8 3.0	3.9 3.0	56.7
57.0	4.1 3.1	4.1 3.1	4.1 3.1	4.1 3.1	4.2 3.1	4.2 3.2	4.2 3.2	57.0
57.3	4.5 3.3	4.5 3.3	4.5 3.3	4.5 3.3	4.5 3.3	4.6 3.4	4.6 3.4	57.3
57.6	4.9 3.5	4.9 3.5	4.9 3.5	4.9 3.5	4.9 3.5	4.9 3.6	4.9 3.6	57.6
57.9	5.3 3.8	5.3 3.8	5.2 3.8	5.2 3.7	5.2 3.7	5.2 3.7	5.2 3.7	57.9
58.2	5.6 4.0	5.6 4.0	5.6 4.0	5.6 4.0	5.6 3.9	5.6 3.9	5.6 3.9	58.2
58.5	6.0 4.2	6.0 4.2	6.0 4.2	6.0 4.2	6.0 4.1	5.9 4.1	5.9 4.1	58.5
58.8	6.4 4.4	6.4 4.4	6.4 4.4	6.3 4.4	6.3 4.3	6.3 4.3	6.2 4.2	58.8
59.1	6.8 4.6	6.8 4.6	6.7 4.6	6.7 4.6	6.7 4.5	6.6 4.5	6.6 4.4	59.1
59.4	7.2 4.8	7.1 4.8	7.1 4.8	7.1 4.8	7.0 4.7	7.0 4.7	6.9 4.6	59.4
59.7	7.5 5.1	7.5 5.0	7.5 5.0	7.5 5.0	7.4 4.9	7.3 4.8	7.2 4.7	59.7
60.0	7.9 5.3	7.9 5.3	7.9 5.2	7.8 5.2	7.8 5.1	7.7 5.0	7.6 4.9	60.0
60.3	8.3 5.5	8.3 5.5	8.2 5.4	8.2 5.4	8.1 5.3	8.0 5.2	7.9 5.1	60.3
60.6	8.7 5.7	8.7 5.7	8.6 5.7	8.6 5.6	8.5 5.5	8.4 5.4	8.2 5.3	60.6
60.9	9.1 5.9	9.0 5.9	9.0 5.9	8.9 5.8	8.8 5.7	8.7 5.6	8.6 5.4	60.9
61.2	9.5 6.2	9.4 6.1	9.4 6.1	9.3 6.0	9.2 5.9	9.1 5.8	8.9 5.6	61.2
61.5	9.8 6.4	9.8 6.3	9.7 6.3	9.7 6.2	9.5 6.1	9.4 5.9	9.2 5.8	61.5

DIP

Ht. of Eye (m)	Ht. of Eye (ft)	Corrn	Ht. of Eye (m)	Ht. of Eye (ft)	Corrn
2.4	8.0	−2.8	9.5	31.5	−5.5
2.6	8.6	−2.9	9.9	32.7	−5.6
2.8	9.2	−3.0	10.3	33.9	−5.7
3.0	9.8	−3.1	10.6	35.1	−5.8
3.2	10.5	−3.2	11.0	36.3	−5.9
3.4	11.2	−3.3	11.4	37.6	−6.0
3.6	11.9	−3.4	11.8	38.9	−6.1
3.8	12.6	−3.5	12.2	40.1	−6.2
4.0	13.3	−3.6	12.6	41.5	−6.3
4.3	14.1	−3.7	13.0	42.8	−6.4
4.5	14.9	−3.8	13.4	44.2	−6.5
4.7	15.7	−3.9	13.8	45.5	−6.6
5.0	16.5	−4.0	14.2	46.9	−6.7
5.2	17.4	−4.1	14.7	48.4	−6.8
5.5	18.3	−4.2	15.1	49.8	−6.9
5.8	19.1	−4.3	15.5	51.3	−7.0
6.1	20.1	−4.4	16.0	52.8	−7.1
6.3	21.0	−4.5	16.5	54.3	−7.2
6.6	22.0	−4.6	16.9	55.8	−7.3
6.9	22.9	−4.7	17.4	57.4	−7.4
7.2	23.9	−4.8	17.9	58.9	−7.5
7.5	24.9	−4.9	18.4	60.5	−7.6
7.9	26.0	−5.0	18.8	62.1	−7.7
8.2	27.1	−5.1	19.3	63.8	−7.8
8.5	28.1	−5.2	19.8	65.4	−7.9
8.8	29.2	−5.3	20.4	67.1	−8.0
9.2	30.4	−5.4	20.9	68.8	−8.1
9.5	31.5		21.4	70.5	

MOON CORRECTION TABLE

The correction is in two parts; the first correction is taken from the upper part of the table with argument apparent altitude, and the second from the lower part, with argument H.P., in the same column as that from which the first correction was taken. Separate corrections are given in the lower part for lower (L) and upper (U) limbs. All corrections are to be **added** to apparent altitude, *but 30' is to be subtracted from the altitude of the upper limb.*

For corrections for pressure and temperature see page A4.

For bubble sextant observations ignore dip, take the mean of upper and lower limb corrections and subtract 15' from the altitude.

App. Alt. = Apparent altitude = Sextant altitude corrected for index error and dip.

These pages are reproduced from *The Nautical Almanac*

Appendix E (continued)

ALTITUDE CORRECTION TABLES 35°–90°—MOON

App. Alt.	35°–39° Corrⁿ	40°–44° Corrⁿ	45°–49° Corrⁿ	50°–54° Corrⁿ	55°–59° Corrⁿ	60°–64° Corrⁿ	65°–69° Corrⁿ	70°–74° Corrⁿ	75°–79° Corrⁿ	80°–84° Corrⁿ	85°–89° Corrⁿ	App. Alt.
00	35 56.5	40 53.7	45 50.5	50 46.9	55 43.1	60 38.9	65 34.6	70 30.1	75 25.3	80 20.5	85 15.6	00
10	56.4	53.6	50.4	46.8	42.9	38.8	34.4	29.9	25.2	20.4	15.5	10
20	56.3	53.5	50.2	46.7	42.8	38.7	34.3	29.7	25.0	20.2	15.3	20
30	56.2	53.4	50.1	46.5	42.7	38.5	34.1	29.6	24.9	20.0	15.1	30
40	56.2	53.3	50.0	46.4	42.5	38.4	34.0	29.4	24.7	19.9	15.0	40
50	56.1	53.2	49.9	46.3	42.4	38.2	33.8	29.3	24.5	19.7	14.8	50
00	36 56.0	41 53.1	46 49.8	51 46.2	56 42.3	61 38.1	66 33.7	71 29.1	76 24.4	81 19.6	86 14.6	00
10	55.9	53.0	49.7	46.0	42.1	37.9	33.5	29.0	24.2	19.4	14.5	10
20	55.8	52.8	49.5	45.9	42.0	37.8	33.4	28.8	24.1	19.2	14.3	20
30	55.7	52.7	49.4	45.8	41.8	37.7	33.2	28.7	23.9	19.1	14.1	30
40	55.6	52.6	49.3	45.7	41.7	37.5	33.1	28.5	23.8	18.9	14.0	40
50	55.5	52.5	49.2	45.5	41.6	37.4	32.9	28.3	23.6	18.7	13.8	50
00	37 55.4	42 52.4	47 49.1	52 45.4	57 41.4	62 37.2	67 32.8	72 28.2	77 23.4	82 18.6	87 13.7	00
10	55.3	52.3	49.0	45.3	41.3	37.1	32.6	28.0	23.3	18.4	13.5	10
20	55.2	52.2	48.8	45.2	41.2	36.9	32.5	27.9	23.1	18.2	13.3	20
30	55.1	52.1	48.7	45.0	41.0	36.8	32.3	27.7	22.9	18.1	13.2	30
40	55.0	52.0	48.6	44.9	40.9	36.6	32.2	27.6	22.8	17.9	13.0	40
50	55.0	51.9	48.5	44.8	40.8	36.5	32.0	27.4	22.6	17.8	12.8	50
00	38 54.9	43 51.8	48 48.4	53 44.6	58 40.6	63 36.4	68 31.9	73 27.2	78 22.5	83 17.6	88 12.7	00
10	54.8	51.7	48.2	44.5	40.5	36.2	31.7	27.1	22.3	17.4	12.5	10
20	54.7	51.6	48.1	44.4	40.3	36.1	31.6	26.9	22.1	17.3	12.3	20
30	54.6	51.5	48.0	44.2	40.2	35.9	31.4	26.8	22.0	17.1	12.2	30
40	54.5	51.4	47.9	44.1	40.1	35.8	31.3	26.6	21.8	16.9	12.0	40
50	54.4	51.2	47.8	44.0	39.9	35.6	31.1	26.5	21.7	16.8	11.8	50
00	39 54.3	44 51.1	49 47.6	54 43.9	59 39.8	64 35.5	69 31.0	74 26.3	79 21.5	84 16.6	89 11.7	00
10	54.2	51.0	47.5	43.7	39.6	35.3	30.8	26.1	21.3	16.5	11.5	10
20	54.1	50.9	47.4	43.6	39.5	35.2	30.7	26.0	21.2	16.3	11.4	20
30	54.0	50.8	47.3	43.5	39.4	35.0	30.5	25.8	21.0	16.1	11.2	30
40	53.9	50.7	47.2	43.3	39.2	34.9	30.4	25.7	20.9	16.0	11.0	40
50	53.8	50.6	47.0	43.2	39.1	34.7	30.2	25.5	20.7	15.8	10.9	50

H.P.	L U	L U	L U	L U	L U	L U	L U	L U	L U	L U	L U	H.P.
54.0	1.1 1.7	1.3 1.9	1.5 2.1	1.7 2.4	2.0 2.6	2.3 2.9	2.6 3.2	2.9 3.5	3.2 3.8	3.5 4.1	3.8 4.5	54.0
54.3	1.4 1.8	1.6 2.0	1.8 2.2	2.0 2.5	2.3 2.7	2.5 3.0	2.8 3.2	3.0 3.5	3.3 3.8	3.6 4.1	3.9 4.4	54.3
54.6	1.7 2.0	1.9 2.2	2.1 2.4	2.3 2.6	2.5 2.8	2.7 3.0	3.0 3.3	3.2 3.5	3.5 3.8	3.7 4.1	4.0 4.3	54.6
54.9	2.0 2.2	2.2 2.3	2.3 2.5	2.5 2.7	2.7 2.9	2.9 3.1	3.2 3.3	3.4 3.5	3.6 3.8	3.9 4.0	4.1 4.3	54.9
55.2	2.3 2.3	2.5 2.4	2.6 2.6	2.8 2.8	3.0 2.9	3.2 3.1	3.4 3.3	3.6 3.5	3.8 3.7	4.0 4.0	4.2 4.2	55.2
55.5	2.7 2.5	2.8 2.6	2.9 2.7	3.1 2.9	3.2 3.0	3.4 3.2	3.6 3.4	3.7 3.5	3.9 3.7	4.1 3.9	4.3 4.1	55.5
55.8	3.0 2.6	3.1 2.7	3.2 2.8	3.3 3.0	3.5 3.1	3.6 3.3	3.8 3.4	3.9 3.6	4.1 3.7	4.2 3.9	4.4 4.0	55.8
56.1	3.3 2.8	3.4 2.9	3.5 3.0	3.6 3.1	3.7 3.2	3.8 3.3	4.0 3.4	4.1 3.6	4.2 3.7	4.4 3.8	4.5 4.0	56.1
56.4	3.6 2.9	3.7 3.0	3.8 3.1	3.9 3.2	3.9 3.3	4.0 3.4	4.1 3.5	4.3 3.6	4.4 3.7	4.5 3.8	4.6 3.9	56.4
56.7	3.9 3.1	4.0 3.1	4.1 3.2	4.1 3.3	4.2 3.3	4.3 3.4	4.3 3.5	4.4 3.6	4.5 3.7	4.6 3.8	4.7 3.8	56.7
57.0	4.3 3.2	4.3 3.3	4.3 3.3	4.4 3.4	4.4 3.4	4.5 3.5	4.5 3.5	4.6 3.6	4.7 3.6	4.7 3.7	4.8 3.8	57.0
57.3	4.6 3.4	4.6 3.4	4.6 3.4	4.6 3.5	4.7 3.5	4.7 3.5	4.7 3.6	4.8 3.6	4.8 3.6	4.8 3.7	4.9 3.7	57.3
57.6	4.9 3.6	4.9 3.6	4.9 3.6	4.9 3.6	4.9 3.6	4.9 3.6	4.9 3.6	4.9 3.6	5.0 3.6	5.0 3.6	5.0 3.6	57.6
57.9	5.2 3.7	5.2 3.7	5.2 3.7	5.2 3.7	5.2 3.7	5.1 3.6	5.1 3.6	5.1 3.6	5.1 3.6	5.1 3.6	5.1 3.6	57.9
58.2	5.5 3.9	5.5 3.8	5.5 3.8	5.4 3.8	5.4 3.7	5.4 3.7	5.3 3.7	5.3 3.6	5.2 3.6	5.2 3.5	5.2 3.5	58.2
58.5	5.9 4.0	5.8 4.0	5.8 3.9	5.7 3.9	5.6 3.8	5.6 3.8	5.5 3.7	5.5 3.6	5.4 3.6	5.3 3.5	5.3 3.4	58.5
58.8	6.2 4.2	6.1 4.1	6.0 4.1	6.0 4.0	5.9 3.9	5.8 3.8	5.7 3.7	5.6 3.6	5.5 3.5	5.4 3.5	5.3 3.4	58.8
59.1	6.5 4.3	6.4 4.3	6.3 4.2	6.2 4.1	6.1 4.0	6.0 3.9	5.9 3.8	5.8 3.6	5.7 3.5	5.6 3.4	5.4 3.3	59.1
59.4	6.8 4.5	6.7 4.4	6.6 4.3	6.5 4.2	6.4 4.1	6.2 3.9	6.1 3.8	6.0 3.7	5.8 3.5	5.7 3.4	5.5 3.2	59.4
59.7	7.1 4.6	7.0 4.5	6.9 4.4	6.8 4.3	6.6 4.1	6.5 4.0	6.3 3.8	6.2 3.7	6.0 3.5	5.8 3.3	5.6 3.2	59.7
60.0	7.5 4.8	7.3 4.7	7.2 4.5	7.0 4.4	6.9 4.2	6.7 4.0	6.5 3.9	6.3 3.7	6.1 3.5	5.9 3.3	5.7 3.1	60.0
60.3	7.8 5.0	7.6 4.8	7.5 4.7	7.3 4.5	7.1 4.3	6.9 4.1	6.7 3.9	6.5 3.7	6.3 3.5	6.0 3.2	5.8 3.0	60.3
60.6	8.1 5.1	7.9 5.0	7.7 4.8	7.6 4.6	7.3 4.4	7.1 4.2	6.9 3.9	6.7 3.7	6.4 3.4	6.2 3.2	5.9 2.9	60.6
60.9	8.4 5.3	8.2 5.1	8.0 4.9	7.8 4.7	7.6 4.5	7.3 4.2	7.1 4.0	6.8 3.7	6.6 3.4	6.3 3.2	6.0 2.9	60.9
61.2	8.7 5.4	8.5 5.2	8.3 5.0	8.1 4.8	7.8 4.5	7.6 4.3	7.3 4.0	7.0 3.7	6.7 3.4	6.4 3.1	6.1 2.8	61.2
61.5	9.1 5.6	8.8 5.4	8.6 5.1	8.3 4.9	8.1 4.6	7.8 4.3	7.5 4.0	7.2 3.7	6.9 3.4	6.5 3.1	6.2 2.7	61.5

Appendix F

LAT 52°

DECLINATION (15°–29°) CONTRARY NAME TO LATITUDE

N. Lat. {LHA greater than 180° ...... Zn=Z
         {LHA less than 180° ...... Zn=360–Z

| 15° | 16° | 17° | 18° | 19° | 20° | 21° | 22° | 23° | 24° | 25° | 26° | 27° | 28° | 29° |

This table (reduced) is reproduced from A.P. 3270, *Sight Reduction Tables for Air Navigation*, Vol. 3

Appendix G

LAT 40°S

LHA ♈	Hc	Zn	Hc	Zn	Hc	Zn	Hc	Zn	Hc	Zn	Hc	Zn	Hc	Zn
	Alpheratz		*Hamal		RIGEL		*CANOPUS		Peacock		*Nunki		Enif	
0	21 03	002	20 15	031	14 15	089	27 39	137	51 34	226	26 20	257	30 59	320
1	21 04	001	20 38	030	15 01	088	28 10	137	51 01	226	25 36	256	30 29	319
2	21 04	000	21 01	029	15 47	088	28 42	136	50 28	226	24 51	256	29 58	318
3	21 04	359	21 23	028	16 33	087	29 14	136	49 56	226	24 07	255	29 27	317
4	21 03	358	21 44	027	17 19	086	29 46	136	49 23	226	23 22	254	28 55	316
5	21 00	357	22 04	026	18 05	086	30 18	135	48 50	226	22 38	254	28 23	315
6	20 58	356	22 24	025	18 50	085	30 50	135	48 17	226	21 54	253	27 50	314
7	20 54	355	22 44	024	19 36	084	31 23	135	47 44	225	21 10	253	27 17	313
8	20 50	354	23 02	023	20 22	084	31 56	134	47 12	225	20 26	252	26 43	312
9	20 45	353	23 20	022	21 07	083	32 28	134	46 39	225	19 43	252	26 08	311
10	20 39	352	23 37	021	21 53	082	33 02	134	46 06	225	18 59	251	25 34	310
11	20 32	351	23 53	020	22 38	082	33 35	134	45 34	225	18 16	251	24 58	309
12	20 25	350	24 09	019	23 24	081	34 08	133	45 01	225	17 32	250	24 22	309
13	20 17	349	24 24	019	24 09	080	34 42	133	44 29	225	16 49	249	23 46	308
14	20 08	349	24 38	018	24 54	079	35 15	133	43 56	225	16 06	249	23 10	307
	*Hamal		ALDEBARAN		RIGEL		*CANOPUS		Peacock		*FOMALHAUT		Alpheratz	
15	24 52	017	14 45	053	25 40	079	35 49	132	43 24	225	62 46	282	19 59	348
16	25 05	016	15 21	052	26 25	078	36 23	132	42 51	225	62 01	281	19 48	347
17	25 17	015	15 57	051	27 10	077	36 57	132	42 19	224	61 16	280	19 37	346
18	25 28	014	16 33	051	27 54	077	37 32	132	41 47	224	60 31	280	19 26	345
19	25 38	013	17 08	050	28 39	076	38 06	131	41 15	224	59 45	279	19 14	344
20	25 48	012	17 43	049	29 23	075	38 41	131	40 43	224	59 00	278	19 01	343
21	25 57	011	18 18	048	30 08	074	39 15	131	40 11	224	58 14	277	18 47	342
22	26 05	010	18 52	047	30 52	074	39 50	131	39 39	224	57 29	276	18 32	341
23	26 12	009	19 25	047	31 36	073	40 25	130	39 08	223	56 43	276	18 17	340
24	26 18	008	19 58	046	32 20	072	41 00	130	38 36	223	55 57	275	18 02	340
25	26 24	007	20 31	045	33 03	071	41 35	130	38 05	223	55 11	274	17 45	339
26	26 29	006	21 03	044	33 47	070	42 10	130	37 33	223	54 25	273	17 28	338
27	26 33	004	21 35	043	34 30	070	42 46	130	37 02	223	53 40	273	17 11	337
28	26 36	003	22 06	042	35 13	069	43 21	129	36 31	222	52 54	272	16 52	336
29	26 38	002	22 37	041	35 55	068	43 57	129	36 00	222	52 08	271	16 33	335
	Hamal		*ALDEBARAN		RIGEL		SIRIUS		*CANOPUS		Peacock		*FOMALHAUT	
30	26 40	001	23 07	041	36 38	067	25 05	091	44 32	129	35 29	222	51 22	271
31	26 41	000	23 36	040	37 20	066	25 51	091	45 08	129	34 59	222	50 36	270
32	26 41	359	24 05	039	38 02	065	26 37	090	45 44	129	34 28	221	49 50	269
33	26 40	358	24 34	038	38 44	064	27 23	089	46 20	129	33 58	221	49 04	269
34	26 38	357	25 02	037	39 25	063	28 09	089	46 56	129	33 28	221	48 18	268
35	26 35	356	25 29	036	40 06	062	28 55	088	47 32	128	32 58	221	47 32	268
36	26 32	355	25 56	035	40 46	062	29 40	087	48 08	128	32 28	220	46 46	267
37	26 28	354	26 22	034	41 27	061	30 26	087	48 44	128	31 58	220	46 00	266
38	26 23	353	26 47	033	42 06	060	31 12	086	49 20	128	31 29	220	45 14	266
39	26 17	352	27 12	032	42 46	059	31 58	085	49 56	128	30 59	220	44 29	265
40	26 10	351	27 36	031	43 25	058	32 44	085	50 33	128	30 30	219	43 43	265
41	26 03	350	28 00	030	44 03	057	33 30	084	51 09	128	30 01	219	42 57	264
42	25 54	349	28 22	029	44 41	055	34 15	083	51 45	128	29 32	219	42 11	263
43	25 45	348	28 44	028	45 19	054	35 01	082	52 21	128	29 03	218	41 26	263
44	25 36	347	29 06	027	45 56	053	35 46	082	52 58	128	28 35	218	40 40	262
	*ALDEBARAN		RIGEL		SIRIUS		*CANOPUS		ACHERNAR		*FOMALHAUT		Hamal	
45	29 26	026	46 33	052	36 32	081	53 34	128	68 01	211	39 55	262	25 25	346
46	29 46	025	47 09	051	37 17	080	54 11	128	67 37	212	39 09	261	25 14	345
47	30 05	024	47 44	050	38 02	080	54 47	128	67 12	213	38 24	261	25 01	344
48	30 24	023	48 19	049	38 48	079	55 23	128	66 47	214	37 38	260	24 48	343
49	30 41	022	48 53	047	39 33	078	56 00	128	66 22	214	36 53	260	24 35	342
50	30 58	021	49 27	046	40 17	077	56 36	128	65 55	215	36 08	259	24 20	341
51	31 14	020	49 59	045	41 02	076	57 12	128	65 29	216	35 23	259	24 05	340
52	31 29	019	50 31	044	41 47	076	57 49	128	65 01	217	34 38	258	23 49	339
53	31 43	018	51 03	042	42 31	075	58 25	128	64 34	217	33 53	257	23 33	338
54	31 57	016	51 33	041	43 16	074	59 01	128	64 06	218	33 08	257	23 15	337
55	32 09	015	52 03	039	44 00	073	59 37	128	63 37	218	32 24	256	22 57	336
56	32 21	014	52 32	038	44 44	072	60 13	129	63 09	219	31 39	256	22 39	336
57	32 32	013	53 00	037	45 27	071	60 49	129	62 39	220	30 54	255	22 19	335
58	32 42	012	53 26	035	46 11	071	61 24	129	62 10	220	30 10	255	21 59	334
59	32 51	011	53 52	034	46 54	070	62 00	129	61 40	220	29 26	254	21 38	333
	ALDEBARAN		*BETELGEUSE		SIRIUS		*ACRUX		ACHERNAR		*FOMALHAUT		Hamal	
60	32 59	010	35 50	036	47 37	069	21 30	157	61 10	221	28 42	254	21 17	332
61	33 07	009	36 17	034	48 20	068	21 48	156	60 40	221	27 57	253	20 55	331
62	33 13	008	36 42	033	49 02	067	22 06	156	60 10	222	27 14	253	20 32	330
63	33 19	006	37 07	032	49 44	066	22 25	156	59 39	222	26 30	252	20 09	329
64	33 23	005	37 31	031	50 26	065	22 44	155	59 08	222	25 46	252	19 45	328
65	33 27	004	37 55	030	51 07	064	23 03	155	58 37	223	25 02	251	19 21	327
66	33 30	003	38 17	029	51 48	063	23 23	155	58 06	223	24 19	251	18 56	327
67	33 32	002	38 39	028	52 29	061	23 43	154	57 35	223	23 36	250	18 30	326
68	33 33	001	39 00	026	53 09	060	24 03	154	57 03	223	22 53	250	18 04	325
69	33 33	359	39 20	025	53 49	059	24 23	154	56 32	224	22 10	249	17 37	324
70	33 32	358	39 39	024	54 28	058	24 44	153	56 00	224	21 27	248	17 10	323
71	33 30	357	39 57	023	55 06	057	25 05	153	55 28	224	20 44	248	16 42	322
72	33 27	356	40 14	021	55 45	055	25 26	153	54 56	224	20 02	247	16 13	321
73	33 24	355	40 31	020	56 22	054	25 47	152	54 24	224	19 19	247	15 45	321
74	33 19	354	40 46	019	56 59	053	26 08	152	53 52	224	18 37	246	15 15	320
	BETELGEUSE		*SIRIUS		Suhail		*ACRUX		ACHERNAR		*Diphda		ALDEBARAN	
75	41 01	018	57 35	051	44 50	115	26 30	152	53 20	224	30 55	272	33 14	353
76	41 14	016	58 11	050	45 31	115	26 52	151	52 48	224	30 09	271	33 07	351
77	41 27	015	58 46	049	46 13	115	27 14	151	52 15	225	29 23	270	33 00	350
78	41 38	014	59 20	047	46 55	114	27 36	151	51 43	225	28 37	270	32 52	349
79	41 49	013	59 53	046	47 37	114	27 59	151	51 11	225	27 51	269	32 43	348
80	41 58	011	60 26	044	48 19	114	28 21	150	50 39	225	27 06	268	32 33	347
81	42 07	010	60 57	042	49 01	114	28 44	150	50 06	225	26 20	268	32 22	346
82	42 14	009	61 27	041	49 43	113	29 07	150	49 34	225	25 34	267	32 11	345
83	42 20	007	61 57	039	50 25	113	29 31	149	49 02	225	24 48	267	31 58	344
84	42 25	006	62 25	037	51 08	113	29 54	149	48 29	225	24 02	266	31 45	343
85	42 30	005	62 52	035	51 50	112	30 18	149	47 57	225	23 16	265	31 30	341
86	42 33	003	63 18	033	52 33	112	30 42	149	47 25	225	22 30	265	31 15	340
87	42 35	002	63 43	031	53 15	112	31 06	148	46 52	225	21 45	264	30 59	339
88	42 36	001	64 06	029	53 58	112	31 30	148	46 20	224	20 59	263	30 43	338
89	42 36	359	64 28	027	54 41	111	31 54	148	45 48	224	20 13	263	30 25	337

LAT 40°S

LHA ♈	Hc	Zn	Hc	Zn	Hc	Zn	Hc	Zn	Hc	Zn	Hc	Zn	Hc	Zn
	PROCYON		REGULUS		*Suhail		ACRUX		*ACHERNAR		RIGEL		*BETELGEUSE	
90	39 25	032	12 43	062	55 23	111	32 19	148	45 16	224	56 33	339	42 34	358
91	39 50	031	13 24	061	56 06	111	32 44	147	44 44	224	56 16	337	42 32	356
92	40 13	030	14 04	060	56 49	111	33 08	147	44 12	224	55 57	335	42 29	355
93	40 35	029	14 44	060	57 32	110	33 33	147	43 40	224	55 37	334	42 24	354
94	40 57	027	15 23	059	58 15	110	33 58	147	43 08	224	55 16	332	42 19	352
95	41 18	026	16 03	058	58 59	110	34 24	147	42 36	224	54 54	330	42 12	351
96	41 37	025	16 42	057	59 42	110	34 49	146	42 04	224	54 31	329	42 05	350
97	41 56	024	17 20	057	60 25	110	35 15	146	41 33	223	54 06	327	41 56	348
98	42 14	022	17 58	056	61 08	109	35 40	146	41 01	223	53 41	326	41 46	347
99	42 31	021	18 36	055	61 52	109	36 06	146	40 30	223	53 14	324	41 36	346
100	42 47	020	19 14	054	62 35	109	36 32	146	39 58	223	52 47	323	41 24	345
101	43 02	018	19 51	054	63 19	109	36 58	145	39 27	223	52 19	321	41 11	343
102	43 16	017	20 28	053	64 02	109	37 24	145	38 56	223	51 50	320	40 57	342
103	43 29	016	21 04	052	64 46	109	37 51	145	38 25	222	51 20	319	40 43	341
104	43 41	014	21 40	051	65 29	109	38 17	145	37 54	222	50 49	317	40 27	339
	PROCYON		REGULUS		*Gienah		ACRUX		*ACHERNAR		RIGEL		*BETELGEUSE	
105	43 52	013	22 16	050	19 41	097	38 43	145	37 23	222	50 17	316	40 11	338
106	44 02	012	22 51	049	20 27	096	39 10	145	36 52	222	49 45	315	39 53	337
107	44 11	010	23 25	049	21 13	095	39 36	145	36 22	222	49 12	313	39 35	336
108	44 19	009	24 00	048	21 59	095	40 03	144	35 51	221	48 38	312	39 15	335
109	44 25	008	24 33	047	22 44	094	40 30	144	35 21	221	48 03	311	38 55	333
110	44 31	006	25 07	046	23 30	093	40 57	144	34 51	221	47 28	310	38 34	332
111	44 35	005	25 39	045	24 16	093	41 24	144	34 21	221	46 53	308	38 12	331
112	44 38	003	26 12	044	25 02	092	41 51	144	33 51	220	46 16	307	37 49	330
113	44 41	002	26 43	043	25 48	092	42 18	144	33 21	220	45 39	306	37 26	329
114	44 42	001	27 15	042	26 34	091	42 45	144	32 51	220	45 02	305	37 02	328
115	44 42	359	27 45	041	27 20	090	43 12	144	32 22	220	44 24	304	36 37	326
116	44 40	358	28 15	040	28 06	090	43 39	144	31 53	219	43 46	303	36 11	325
117	44 38	356	28 45	039	28 52	089	44 06	144	31 24	219	43 07	302	35 44	324
118	44 35	355	29 14	038	29 38	088	44 34	144	30 55	219	42 28	301	35 17	323
119	44 30	354	29 42	037	30 24	088	45 01	144	30 26	219	41 48	300	34 49	322
	REGULUS		*SPICA		ACRUX		*CANOPUS		RIGEL		BETELGEUSE		*PROCYON	
120	30 09	036	13 59	093	45 28	144	69 12	224	41 08	299	34 20	321	44 24	352
121	30 36	035	14 45	092	45 56	144	68 40	225	40 28	298	33 51	320	44 18	351
122	31 03	034	15 31	091	46 23	144	68 07	226	39 47	297	33 21	319	44 10	349
123	31 28	033	16 17	091	46 50	144	67 34	227	39 06	296	32 51	318	44 01	348
124	31 53	032	17 03	090	47 17	144	67 01	227	38 25	295	32 20	317	43 51	347
125	32 18	031	17 49	090	47 45	144	66 27	228	37 43	294	31 48	316	43 40	345
126	32 41	030	18 35	089	48 12	144	65 53	228	37 01	293	31 16	315	43 28	344
127	33 04	029	19 21	088	48 39	144	65 18	229	36 19	293	30 43	314	43 15	343
128	33 26	028	20 06	088	49 06	144	64 44	229	35 36	292	30 10	313	43 00	341
129	33 47	027	20 52	087	49 33	144	64 09	230	34 53	291	29 36	312	42 45	340
130	34 08	026	21 38	086	50 01	144	63 34	230	34 10	290	29 01	311	42 29	339
131	34 27	025	22 24	086	50 28	144	62 58	230	33 27	289	28 26	310	42 12	337
132	34 46	024	23 10	085	50 55	144	62 23	231	32 44	288	27 51	309	41 54	336
133	35 04	023	23 56	084	51 21	144	61 47	231	32 00	288	27 15	308	41 35	335
134	35 21	021	24 41	084	51 48	144	61 12	231	31 16	287	26 39	308	41 15	334
	*REGULUS		SPICA		*ACRUX		CANOPUS		*RIGEL		BETELGEUSE		PROCYON	
135	35 38	020	25 27	083	52 15	145	60 36	231	30 32	286	26 02	307	40 54	332
136	35 53	019	26 13	082	52 41	145	60 00	231	29 48	285	25 25	306	40 32	331
137	36 08	018	26 58	082	53 08	145	59 24	232	29 03	285	24 48	305	40 10	330
138	36 21	017	27 44	081	53 34	145	58 48	232	28 19	284	24 10	304	39 46	329
139	36 34	016	28 29	080	54 00	145	58 12	232	27 34	283	23 32	303	39 22	328
140	36 46	014	29 14	079	54 26	146	57 36	232	26 49	282	22 53	302	38 57	326
141	36 57	013	29 59	079	54 52	146	56 59	232	26 04	282	22 14	302	38 31	325
142	37 07	012	30 44	078	55 18	146	56 23	232	25 19	281	21 35	301	38 05	324
143	37 16	011	31 29	077	55 44	146	55 47	232	24 34	280	20 55	300	37 38	323
144	37 24	009	32 14	076	56 09	147	55 10	232	23 49	280	20 15	299	37 10	322
145	37 31	008	32 58	076	56 34	147	54 34	232	23 03	279	19 35	298	36 41	321
146	37 37	007	33 43	075	56 59	147	53 58	232	22 18	278	18 55	298	36 12	320
147	37 42	006	34 27	074	57 24	148	53 21	232	21 32	277	18 14	297	35 41	319
148	37 46	005	35 11	073	57 48	148	52 45	232	20 47	277	17 33	296	35 11	318
149	37 50	003	35 55	073	58 12	149	52 08	232	20 01	276	16 51	295	34 40	317
	REGULUS		*SPICA		ACRUX		*Miaplacidus		CANOPUS		SIRIUS		*PROCYON	
150	37 52	002	36 39	072	58 36	149	59 46	188	51 32	232	41 43	284	34 08	316
151	37 53	001	37 23	071	59 00	149	59 39	189	50 56	232	40 58	283	33 35	315
152	37 53	000	38 06	070	59 23	150	59 32	189	50 20	232	40 14	283	33 02	314
153	37 52	358	38 49	069	59 46	150	59 24	190	49 43	232	39 29	282	32 29	313
154	37 50	357	39 32	068	60 08	151	59 16	191	49 07	232	38 44	281	31 55	312
155	37 48	356	40 14	067	60 31	151	59 07	191	48 31	232	37 58	280	31 20	311
156	37 44	355	40 57	067	60 53	152	58 57	192	47 55	232	37 13	280	30 45	310
157	37 39	353	41 39	066	61 14	153	58 48	193	47 19	232	36 28	279	30 09	309
158	37 33	352	42 20	065	61 35	153	58 38	193	46 43	231	35 42	278	29 33	308
159	37 27	351	43 02	064	61 56	154	58 27	194	46 07	231	34 57	277	28 57	307
160	37 19	350	43 43	063	62 16	154	58 16	194	45 31	231	34 11	277	28 20	306
161	37 10	349	44 24	062	62 35	155	58 04	195	44 55	231	33 26	276	27 43	305
162	37 01	347	45 04	061	62 54	156	57 52	195	44 20	231	32 40	275	27 05	304
163	36 50	346	45 44	060	63 13	157	57 40	196	43 44	231	31 54	275	26 27	304
164	36 39	345	46 23	059	63 31	157	57 27	196	43 09	230	31 08	274	25 48	303
	ARCTURUS		*ANTARES		ACRUX		*CANOPUS		SIRIUS		PROCYON		*REGULUS	
165	15 24	047	22 29	106	63 48	158	42 33	230	30 22	273	25 09	302	36 26	344
166	15 57	046	23 13	106	64 05	159	41 58	230	29 36	273	24 30	301	36 13	343
167	16 30	046	23 57	105	64 21	160	41 23	230	28 51	272	23 51	300	35 59	341
168	17 03	045	24 41	105	64 37	161	40 48	230	28 05	271	23 11	299	35 44	340
169	17 35	044	25 26	104	64 52	161	40 13	229	27 19	271	22 30	299	35 28	339
170	18 07	043	26 11	104	65 06	162	39 38	229	26 33	270	21 50	298	35 11	338
171	18 38	042	26 55	103	65 20	163	39 03	229	25 47	269	21 09	297	34 53	337
172	19 08	042	27 40	102	65 33	164	38 28	229	25 01	269	20 28	296	34 35	336
173	19 39	041	28 25	102	65 45	165	37 54	229	24 15	268	19 47	295	34 15	335
174	20 08	040	29 10	101	65 56	166	37 19	228	23 29	267	19 05	295	33 55	333
175	20 38	039	29 55	101	66 06	167	36 45	228	22 43	267	18 23	294	33 34	332
176	21 06	038	30 40	100	66 16	168	36 11	228	21 57	266	17 41	293	33 13	331
177	21 34	037	31 26	100	66 25	169	35 37	227	21 11	266	16 59	292	32 50	330
178	22 02	036	32 11	099	66 33	171	35 03	227	20 25	265	16 16	292	32 27	329
179	22 29	035	32 57	098	66 40	172	34 30	227	19 40	264	15 33	291	32 03	328

This table is reproduced from A.P. 3270, *Sight Reduction Tables for Air Navigation*, Vol. 1

Appendix H

TABLE 5.—Correction to Tabulated Altitude for Minutes of Declination

d / '	1	2	3	4	5	6	7	8	9	10	11	12	13	14	15	16	17	18	19	20	21	22	23	24	25	26	27	28	29	30	31	32	33	34	35	36	37	38	39	40	41	42	43	44	45	46	47	48	49	50	51	52	53	54	55	56	57	58	59	60	d / '
0	0	0	0	0	0	0	0	0	0	0	0	0	0	0	0	0	0	0	0	0	0	0	0	0	0	0	0	0	0	0	0	0	0	0	0	0	0	0	0	0	0	0	0	0	0	0	0	0	0	0	0	0	0	0	0	0	0	0	0	0	0

This table is reproduced from A.P. 3270, *Sight Reduction Tables for Air Navigation*, Vol. 3

INDEX

Admiralty chart 50
Air Almanac 21
Aldebaran 44
★*Alkaid* 38
Almanacs 10, 20, 21
Altair 38
Altitude 7
  apparent 27
  calculated 16
  correction 34
  parallax in 20
  sextant 19, 32, 41
    correction 26–27
  tables 22, 48
  tabulated 16–19, 27
    correction 30
  true 17–19, 29, 32
Altitude Correction Tables 27, 41, 50
★*Antares* 38, 44
Apparent altitude 27
Arcturus 38
Aries 35–36, 40
Assumed latitude 25, 26
Assumed position 16, 25–26
Azimuth 8–9, 29–30, 42
  tables 22, 48
Azimuth angle 8–9, 17, 18, 27, 29
Azimuth arrows 41–42

Calculated altitude 16
Canopus 44
Chosen position 16
Civil twilight 37
Contrary name 14, 27
Correction 19–20
  altitude 34
  declination 26
  dip 19
  double second difference 50
  refraction 19
  semi-diameter 19–20

Correction *continued*
  sextant altitude 26–27
  tabulated altitude 30

Deck-watch time 24
Declination 1–2, 26–29
  correction 26
  of heavenly bodies 2
  of Moon 2
  of stars 37
  of Sun 2
Declination increment 28–29
★*Deneb* 38
Dip 32, 33, 41
  correction for 19
Double second difference, correction for
  50

Elevated Pole 8
Ephemerides 21
Equation of Time 10

First Point of Aries 35
Fomalhaut 44

Geographical position 1
Great circles 9–10
Greenwich Hour Angle 3
Greenwich Mean Time 10

Hamal 38, 44
Heavenly bodies 1
  declinations of 2
Horizon 6–7
  artificial 50
  clarity of 23
Horizontal parallax 20
Hour angle 2–6

Index error 32, 33, 41, 47, 50
Intercept 17
Intercept rule 18

Index

Jupiter 32

Kochab 38

Latitude, assumed 25, 26
Local Hour Angle 3
   Sun from GMT 24–25

Marine Sight Reduction Tables 49
Mars 32
Mercator plotting sheets 51
Meridian passage 6, 40
Meridian sights 34–35
Moon 1
   declination of 2
Moon Altitude Correction Tables 50
Moon sights 31–32, 51

Nautical Almanac 21, 24, 27, 30, 31, 32,
   34, 37
Nautical twilight 37
Noon sight 34
Nunki 38
Nutation 40

Octant 46

Parallax, correction for 20
   horizontal 20
   in altitude 20
Peacock 44
Planet identification 34
Planet sights 32–34
Planets 1
Plotting sheets 50–52
Polaris 13, 14, 40
   see also Pole Star sights
Pole Star sights 40–45
Position, selection of 41–45
Position circle 10, 11
Position line 10–12, 17
   accuracy of 45
   error 41
Precession 40

Reed's Almanac 21
Refraction 41, 50
   correction for 19
Rigel 44

Same name 14, 27
Saturn 32
Semi-diameter, correction 19–20
Sextant 7, 46
Sextant altitude 19, 32, 41
   correction 26–27
Sextant error 46
Side error 47
Sidereal Hour Angle 6, 35
Sight Reduction Tables for Air Navigation
   22
Sight Reduction Tables for Marine
   Navigation 22
Sights, best time to take 12–13
   meridian 34–35
   Moon 31–32, 51
   noon 34
   planet 32–34
   Pole Star 40
   practice 50
   star 35–40
   Sun 22–31
Sirius 36, 40, 44
Spherical triangle 15–19, 22
   basic formulae 48
Star globe 47–48
Star sights 35–40
Stars 1
   declinations of 37
Sun 1
   declination of 2
Sun sights 22–31

Tables 21
   Sight Reduction 22
Tables of Computed Altitude and Azimuth
   22, 48
Tabulated altitude 16–19, 27
   correction 30
True altitude 17–19, 29, 32
Twilight, civil 37
   nautical 37

Venus 32, 33

Zenith 6
Zenith distance 7, 15, 16